ITALY

MAJOR WORLD NATIONS
ITALY

Kathryn Bonomi

CHELSEA HOUSE PUBLISHERS
Philadelphia

FRONTISPIECE: Bernini's Fountain of the Four Rivers in Rome

Chelsea House Publishers

Contributing Author: Christina Schlank

Copyright © 1999 by Chelsea House Publishers,
a division of Main Line Book Co.
All rights reserved.
Printed and bound in the United States of America.

First Printing

1   3   5   7   9   8   6   4   2

Library of Congress Cataloging-in-Publication Data

Bonomi, Kathryn W.
Italy/Kathryn W. Bonomi
p. cm.—(Major world nations)
Summary: Describes the history, geography, government, society, economy,
and culture of Italy.
1. Italy. [1. Italy.] I. Title. II. Series.
DG30.B66   1990   90-1382   CIP AC
ISBN 0-7910-4760-1

CONTENTS

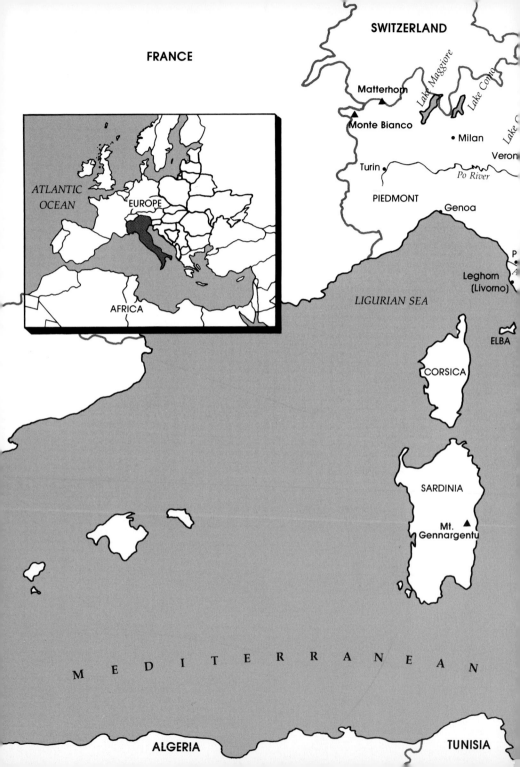

FACTS AT A GLANCE

Land and People

Area	116,303 square miles (301,225 square kilometers)
Highest Point	Monte Bianco (Mont Blanc), 15,771 feet (4,807 meters)
Greatest Length	736 miles (1,178 kilometers)
Greatest Width	391 miles (626 kilometers)
Major Rivers	Po, Adige, Brenta, Reno, Tiber, Arno
Major Lakes	Garda, Maggiore, Como, Bolsena, Bracciano, Trasimeno
Capital	Rome (population 2.7 million)
Other Major Cities	Milan, Naples, Turin, Palermo, Genoa, Bologna, Florence
Population	57.5 million
Population Density	494 people per square mile (191 per square kilometer)
Population Distribution	Rural, 33 percent; urban, 67 percent
Official Language	Italian

Literacy Rate	98 percent
Ethnic Groups	Italian, 98 percent; other, 2 percent (includes Germans, Slovenes, and French)
Religions	Roman Catholic, 83 percent; other, 17 percent (includes 16 percent nonreligious and atheist and 1 percent other religions)
Infant Mortality Rate	8 per 1,000 live births
Average Life Expectancy	82 years for women, 75 years for men

Economy

Chief Exports	Metals, textiles and clothing, machinery, motor vehicles, transportation equipment, chemicals
Chief Imports	Petroleum, industrial machinery, raw materials, food
Chief Agricultural Products	Wheat, barley, oats, rice, corn, beans, wine, olive oil
Industries	Automobile manufacturing, textiles and clothing, electronics
Tourism	55 million tourists in 1996
Currency	lira, divided into 100 centesimi; U.S. $1 equal to approximately 1,700 lire in 1997

Government

Form of Government	Republic with two legislative houses
Formal Head of State	President, elected to seven-year term
Head of Government	Prime Minister, appointed by the president
Legislature	Parliament, composed of the Chamber of Deputies and the Senate

Local Government   20 regions, divided into 94 provinces and
subdivided into 8,091 communes; 5 regions
have semi-independent status; regions and
communes are governed by local councils

HISTORY AT A GLANCE

753 B.C.	According to legend, Rome is founded by Romulus, its first king.
5th century B.C.	The Romans overthrow the Etruscan monarchy and establish their own rule.
264–201 B.C.	The first two Punic Wars between the Romans and the Carthaginians end with the defeat of the Carthaginian leader Hannibal by Scipio at Zama.
49 B.C.	Julius Caesar returns to Rome after conquering Gaul, drives out Pompey, and proclaims himself dictator. On March 15, 44 B.C., he is assassinated by a gang made up of former allies, including Brutus and Cassius.
27 B.C.	Octavian becomes the first Roman emperor and is given the name Augustus Caesar.
313	Emperor Constantine declares religious freedom by publishing the Edict of Milan, which makes Christianity the official religion.
3rd to 6th century	Attacks by barbarians begin. Germanic barbarians—Visigoths, Huns, Vandals, and Ostrogoths—make the first incursions. In 568

the Lombards, led by King Alboin, invade northern Italy and make Pavia their capital.

800   Pope Leo III crowns Charlemagne, the king of the Lombards, emperor.

962   Otto I of Saxony is enthroned as the first Holy Roman Emperor. (The empire will last until 1806, when Francis II renounces the title.)

1215–50   Holy Roman Emperor Frederick II reigns from his Sicilian court.

1309–78   The popes abandon Rome and make Avignon, in southern France, their residence.

14th century   Northern and central Italian cities such as Venice, Florence, and Milan form powerful independent states.

1347–51   The Black Death arrives in Sicily, brought by Genoan merchants returning from trade voyages to the East. The plague kills about a third of the population of Europe.

1494–1559   France and Spain vie for control over the northern Italian states.

1527   Rome is sacked by the Spanish troops of Charles V and Spain gains dominance in Italy.

1713   The signing of the Peace of Utrecht marks the end of the Franco-Spanish conflict over Italy.

1805   Napoléon Bonaparte proclaims himself king of Italy. He is defeated in 1814.

1820–32   Revolutionary outbreaks by the Carbonari and members of other secret groups favoring independence are put down by Austria.

1831   Giuseppe Mazzini founds the Young Italy Association and starts the movement toward national unification that is called the Risorgimento.

1860	Garibaldi and the Red Shirts arrive in Sicily and march north to Naples to unify southern Italy.
1861	Most of Italy is united. The Kingdom of Italy is proclaimed; Victor Emmanuel II is named king.
1871	The remaining independent states are brought into the kingdom and Rome is made its capital.
1915	Italy enters World War I on the side of the Allies.
1922	King Victor Emmanuel III is pressured into making Mussolini Italy's premier after the Black Shirts march on Rome in October.
1940	Italy enters World War II on the side of Germany, and Mussolini declares war on France and England.
1943	Allies land in Sicily and war is waged on Italian soil. Italy surrenders.
1946	The Republic of Italy is established.
1957	The European Economic Community is founded in Rome.
1968	Unrest among university students and workers in the Hot Autumn leads to the formation of left-wing groups such as the Red Brigades.
1978	The terrorist Red Brigades kidnap and murder former prime minister Aldo Moro.
1983	Bettino Craxi becomes the first Socialist prime minister of Italy. His term in office ends with the hijacking of the *Achille Lauro*.
1994	A right-center coalition, the Freedom Alliance, wins a majority in the parliament.
1996	The Olive Tree, a left-center coalition led by Romano Prodi, is victorious in national elections.

Founded in 27 B.C. by the consul Agrippa as a temple to the ancient Roman gods, the Pantheon has survived two millennia of often tumultuous history. It stands in the heart of modern Rome as a reminder of Italy's classical origins.

1

Italy and the World

The remarkable history of Italy has been marked by the heights of human achievement. On the Italian peninsula in southern Europe, the powerful Roman Empire rose and fell. From the Church of St. Peter's in Rome, the capital of present-day Italy, the Christian religion spread throughout Europe and the world. During the Renaissance, illustrious artists such as Leonardo da Vinci and Michelangelo painted and carved masterpieces that set the course of modern art. The astronomer Galileo Galilei and other Italian scientists helped bring about the scientific revolution. Writers such as Dante Alighieri contributed to the world's literary heritage, and the music of composers such as Giuseppe Verdi and Giacomo Puccini is heard in concert halls on every continent. Italy's present position as a world leader in the arts is simply the continuation of a long tradition.

The stones of Italy tell its story. Classical temples, medieval churches, Renaissance villas—buildings that have withstood time and the elements—dot the Italian landscape. They are memorials to the lives of princes and slaves, of dictators and revolutionaries. Today's sophisticated, highly modern Italians live amid these centuries-old monuments.

The Italians have long preserved their history by adapting their ancient heritage to new purposes. Consider, for example, the evolution of uses they have found for the Pantheon, a magnificent domed edifice built in Rome in 27 B.C. The Pantheon was originally a temple celebrating the gods and goddesses of the ancient Roman religion. In the 7th century it was converted into a Christian church and renamed Santa Maria ad Martyres (St. Mary of the Martyrs), in honor of the Christian martyrs whose tombs had been transferred there. In the 16th century it became a burial place and memorial to famous artists and political leaders.

Today the Pantheon remains a focal point of the city: Children play around the fountain in the square that borders it, and tourists sit in nearby outdoor cafés or gather before the Pantheon entrance. On a typical day, these tourists might see Italian boys kicking a soccer ball beneath one of the Pantheon's massive granite columns, a woman sitting on the steps sorting roses, and a vendor setting up a souvenir stand in the square. Italians do not view their art and culture with distant awe, as if it belonged in a museum. Instead, they recognize it as the stuff of daily life.

Italy is strategically located between Europe, Africa, and the Middle East. The northern part of the country is sandwiched between western and eastern Europe, and the rest of the country juts into the Mediterranean Sea. On maps, Italy looks like a high-heeled boot poised to kick the island of Sicily (which is part of Italy). The North African nation of Tunisia is fewer than 100 miles (160 kilometers) from Sicily. At the center of the Mediterranean world, the Italian peninsula was a crossroads of ancient cultures. Greek, Middle Eastern, German, French, Spanish, and North African influences have helped to shape Italy's history and culture.

The issue of national identity has concerned Italians since the fall of the Roman Empire in the 6th century A.D. It is often said that Italy never experienced a true sense of national identity until the dictator Benito Mussolini came to power in 1922. This may be explained by

Once the center of an empire, Rome is now the capital of a nation forged from an amalgam of city-states, petty kingdoms, and territorial holdings. The quest for unification and national identity has shaped the country's modern history.

the history of Italy's fiercely independent regions. For centuries, leading cities such as Genoa and Florence were independent city-states. As each highly competitive city prospered, its region developed distinctive characteristics: Genoa and Venice were known as trading centers, Florence was a center of artistic activity, and Bologna, with one of the world's oldest universities, was a seat of learning.

Other differences between the regions developed as a result of foreign occupation. French, German, Spanish, and North African conquerors controlled various parts of Italy in various eras, and each invading culture left its mark. Italy's present unity has had to accommodate these differences. Each of the 20 regions of modern Italy has its own characteristic customs, cuisine, and culture. The regions preserve these local distinctions, watchful of pressure to conform. Five areas have carried the concept of regionalism so far as to have demanded—and received—a degree of self-rule; and one tiny territory inside a northern province has declared itself the independent republic of San Marino, with its own government, flag, and currency.

Another important concern for Italy today is its place in the world. Since it became a unified and independent nation in the late 19th century, Italy has had to contend with a sluggish and often desperate economy, with enormous debts, and with disastrous experiences in war. The nation's defeat in World War II (1939–45) left its cities in ruins and its people poverty-stricken. In the decades since the war, Italy has developed rapidly into an industrialized nation, but too often Italians feel that their nation is not taken seriously as a world power. Some Italians fear that Italy is viewed by the rest of the world as a tourist attraction, valued only for its glorious ancient history and its artworks, its fine food and wine, its fashions, and its scenic views.

But history has taught Italians how to adapt. In the 1950s and 1960s Italy's economy grew faster than that of any nation on earth with the exception of Japan. The Italy of today is progressive, its economy is robust, and most of its people are fairly prosperous. Italy enjoys the fifth strongest economy in the West, having recently surpassed Great Britain. Its products—ranging from sports cars such as the Alfa Romeo and Maserati to clothing by Benetton and

Armani—are prized throughout the world. This measure of well-being represents an enormous feat for a nation that only a few decades ago listed 24 percent of its population as either "destitute" or "in hardship." It also testifies to the resilience of the extraordinary and indomitable Italian people.

A hiker contemplates the steep green slopes and sheltered valleys of Valle d'Aosta in the northwest. The snow-clad peaks in the background are the Italian Alps, part of the great Alpine mountain chain that runs from France to Slovenia.

2

The Lay of the Land

Northern, or continental, Italy is bordered by France on the west, Switzerland on the northwest, Austria on the northeast, and Slovenia on the east. The peninsula that forms the southern three-quarters of the country is bordered by four seas, each of which is part of the Mediterranean Sea. The Ligurian Sea washes the northwestern shores of the peninsula, where the cities of Genoa and Leghorn (Livorno) are located. The Tyrrhenian Sea is the name given to the waters between the western coast of the peninsula and the islands of Sicily and Sardinia. The Ionian Sea lies between southern Italy and Greece, and farther north the Adriatic Sea separates Italy from the Balkan states (formerly Yugoslavia).

The coastline extends around the peninsula for 4,632 miles (7,411 kilometers). Italy has a total area of 116,303 square miles (301,225 square kilometers), including the islands of Sicily and Sardinia; this makes Italy about three-fourths the size of the state of California in the United States. Two independent states, San Marino and Vatican City, lie within Italian territory.

Eighty percent of Italy is mountainous. The two principal mountain chains are the Apennines and the Alps. The Apennine Mountains run from north to south along the entire peninsula and form a rugged spine that separates the eastern part of the country from the western part. The eastern slopes of the Apennines are steeper than the western slopes and are cut by many swift-flowing streams that empty into the Adriatic Sea. The western slopes are the source of many of western Italy's longer rivers, including the Arno and the Tiber. Thirteen major passes have allowed travelers to cross the Apennines from ancient times to the present.

The highest peak of the Apennine chain is Monte Corno, northeast of Rome; its altitude is 9,560 feet (2,914 meters) above sea level. The Apennines continue into the island of Sicily, where they are called the Madonie Mountains. The highest peak there is Mount Etna, an active volcano, at 11,053 feet (3,369 meters). A 6,017-foot (1,834-meter) peak called Gennargentu dominates the mountainous island of Sardinia.

The Italian Alps are part of the great Alpine mountain range that sweeps across south-central Europe from France through Switzerland and northern Italy to Austria. Three of the Alpine ranges in northwestern and northern Italy are called the Pennines, the Savoy Alps, and the Graian Mountains. These Alpine ranges are the site of three mountain lakes that have long been noted for their scenic beauty: Lake Como, Lake Maggiore, and Lake Garda. Also located in the northeast is Italy's highest point—indeed, the highest point in Europe. This is Monte Bianco, called Mont Blanc in French. It straddles the border between France and Italy and is 15,771 feet (4,807 meters) in altitude. A tunnel that runs beneath it between Courmayeur, Italy, and Chamonix-Mont-Blanc, France, is 7.5 miles (12 kilometers) long; it is one of the world's longest traffic tunnels. Another well-known peak, the Matterhorn (called Monte Cervino in Italian), sits on the border between Italy and Switzerland. It is 14,690 feet (4,478 meters) tall.

The northeastern part of the Alpine chain, near the Slovenian border, is called the Dolomite Mountains because it is chiefly made of a stone called dolomite. The craggy Dolomites rise in dramatic pinnacles above forests and grassy plains and are a favorite recreation spot for hikers and climbers. Their tallest peak is Marmolada, with an altitude of 10,965 feet (3,342 meters).

Climate and Weather

Although it is not a large country, Italy has a great range of climates. Temperature and rainfall vary from region to region, according to latitude, elevation above sea level, and distance from the sea. The mountainous areas can be bitterly cold. The valley of the Po River, in the north, endures chilly winters with long periods of frigid weather and much snow in the mountains; summers are hot and humid, drenched in heavy fog. Most of the peninsula enjoys a milder climate, with winters that are cool and rainy and summers that are hot and dry. The south has very mild winters and scorching summers.

The average annual temperature in the Po River valley is about 55°F (13°C). In the coastal lowlands, the average annual temperature is slightly warmer: 57°F (14°C). Sicily is the warmest region, with an average annual temperature of 64°F (18°C). Sardinia's summers are marked by a hot wind from the south called the *scirocco*. December and January are the coldest months throughout the country, and July and August are the hottest.

The annual rainfall increases from south to north. In the country's driest regions—the southeastern coast and the islands of Sicily and Sardinia—the average annual rainfall is about 20 inches (50 centimeters). The Po River valley receives about 40 inches (100 centimeters) of rain yearly, whereas the Alps and some of the western slopes of the Apennines receive twice that much.

The Geological Story

Italy's landscape is still changing and being reshaped by earthquakes and volcanic eruptions. Vesuvius (near the city of Naples), Mount Etna in Sicily, and Stromboli and Vulcano in the Lipari Islands (near Sicily) are all active volcanoes; Etna is the largest active

Archaeologists uncover the mummified remains of people who were buried by lava and ash when an eruption of Mount Vesuvius destroyed the city of Pompeii in A.D. *79.*

volcano in Europe. In addition, the landscape is dotted with extinct volcanoes. The craters of some of the long-dead volcanoes have formed deep lakes, including Lake Bolsena, Lake Bracciano, and Lake Nemi.

Perhaps the most famous volcanic eruption in the history of the peninsula occurred when Vesuvius blew its top in A.D. 79. This disaster was recorded by a Roman named Pliny the Younger. His uncle had gone to investigate a plume of smoke and found himself in the midst of the disaster: "The houses . . . now rocked from side to side with frequent and violent concussions . . . stones and cinder . . . fell in large showers." When Vesuvius erupted, its streaming rivers of lava buried the Roman towns of Pompeii and Herculaneum, and a traveler to Pompeii today can see the petrified remains of doomed citizens and animals who were engulfed as they tried to escape. Nonetheless, Italians continue to live in the shadow of Vesuvius, their houses rebuilt atop the hardened lava left by past eruptions and their vineyards and orchards climbing the mountain slopes.

The Parts of Italy

Although there are many differences among Italy's many regions, the country can be divided into three major zones: north, central, and south.

The north consists of the Alpine ranges and the Po River valley, also called the Po Plain. Visitors from neighboring countries enter northern Italy by well-worn routes through the Italian Alps, such as the St. Bernard Pass, which has been used for several thousand years. Crossing into Italy, they encounter rolling foothills and Italy's picturesque lake district. Lakes Como, Maggiore, Garda, and Iseo fill depressions that were gouged out of the ground by Alpine glaciers. They feed six small rivers that flow south to the Po Plain.

The excavated ruins of Pompeii have given scholars valuable insights into what cities were like 20 centuries ago. Vesuvius broods in the background, still emitting occasional plumes of steam, ash, or smoke. Its most recent major eruption occurred in 1906.

This plain is a large, broad lowland that stretches all the way across the northern part of the country. It accounts for three-fifths of Italy's lowland area. The plain is watered principally by the Po River, which flows 398 miles (639 kilometers) from the eastern mountain heights to the Adriatic Sea. The Po Plain is the most fertile agricultural area in the country. The poplar, oak, willow, and alder groves that once flourished there have largely been replaced by crops, especially rice, which requires abundant water. Dishes based on *risotto* (rice) are a staple of northern Italian cuisine, as pasta is in the south.

The Po Plain is the site of two of Italy's busiest industrial cities. They are Turin (base of the Fiat car company) and Milan (the textile and fashion capital of Italy). Two of the country's largest and oldest ports are also found in this region: Venice, the northeastern lagoon city and birthplace of the world traveler Marco Polo (1254–1324); and Genoa, in the northwest along the Italian Riviera, birthplace of Christopher Columbus (1451–1506).

The part of Italy best known to foreigners is perhaps the central area, where the city of Florence is located. The lush Mediterranean countryside, familiar as the background to many Renaissance paintings, was described by the American writer Henry James this way: "Cypresses cast straight shadows at [the church's] corners, while in the middle grew a wondrous Italian tangle of wheat and corn, vines and figs, peaches and cabbages, memories and images, anything and everything."

Poets have long praised the haunting beauty of the Campagna di Roma, the romantic and desolate countryside pitted with volcanic lakes that surrounds Rome. East of Rome is the Abruzzi region, a harsh, forbidding land of chilly Apennine peaks.

Southern Italy has been a vacation land for several thousand years. Along the coast south of Rome, wealthy citizens of the Roman Empire built their country villas. A little farther south is the harbor city of Naples, and farther south still is a popular seaside resort area called the Amalfi. But most of the south is poor, its inhabitants eking out a meager living on the scorched land. Only Apulia, the "heel of the boot," which has a thriving agricultural economy, and Sicily, a rich source of minerals and petroleum, fare somewhat better. Sardinians have developed an economy based on agriculture and sheepherding, to which they have added the mining of coal, lead, and zinc in recent years.

Flora and Fauna

Italy has been a cultivated land for centuries, and much of its natural wildlife has been replaced by crops and domesticated animals. Grassy meadows, beeches, and conifers thrive along the high mountain slopes, forests of oak and chestnut at lower altitudes. The Alps are roamed by chamois (a type of antelope), ibex (wild goats), and deer. Bears, otters, and wildcats find shelter among the oak and

The lynx, also called a wildcat, lives in the forested areas of the Apennines. It has thick, soft fur, a short tail, tufted ears, and large padded feet and hunts for food at night.

The sun-scorched landscape near Palermo, Sicily, is dotted with olive trees. This rugged island has Italy's hottest and driest climate.

chestnut trees of the Apennine Mountains. Wolves still range parts of the south.

Along the coastline the Mediterranean Sea moderates the climate. At an earlier time cork and holm oak were widely scattered there; now cypress, umbrella pine, juniper, myrtle, thyme, laurel, wheat, olives, and vines flourish in their place. The southern climate and soil support cotton, citrus fruits, vines, and almond trees.

More than 400 species of birds can be spotted in Italy; among the most common are the raven and swallow. The Adriatic Sea is the source of abundant sea life, including perch, mullet, rays, sharks, sardines, anchovies, mackerel, swordfish, and tuna. Sponges and coral are also plentiful.

The Environment

Italy today faces several severe environmental problems. Two of the most serious involve water: They are periodic flooding and the pollution of waterways caused by the runoff from fertilizers, pesticides, and industrial waste.

One of the worst floods in Italy's history occurred on November 4, 1966, when a storm on the Adriatic Sea elevated the water level by more than six feet (two meters), overflowing the banks of the port city of Venice and greatly damaging its precious art. Since that time, several programs have been set under way to control or give warning of floods along the Adriatic coast. Venice faces the additional problem of sinkage. The city is built on pilings sunk into the seabed, and it is sinking at a rate of about .16 inches (4 millimeters) a year. The Italian government has worked for some years on a plan to halt the sinkage, but so far no significant action has been taken.

Another environmental problem is soil erosion, which has been caused by centuries of farming and goat grazing. Air pollution is also a concern. Emissions from cars and factories pollute the air, especially in the northern cities. In Milan, local officials have had to call a citywide emergency several times because of severe air pollution. This type of pollution is particularly harmful to anything made of marble, such as statues and buildings, because contact with certain gases turns marble to lime, a soft substance that can be washed away by rain.

In recent years the government has been active in looking for solutions to Italy's ecological and environmental problems. The National Research Council and the Ministry of Culture and Environmental Quality have been assigned this difficult task. National wildlife preserves have been created—in Gran Paradiso, Stelvio, Circeo, and Abruzzi parks—and hundreds of other areas have been zoned to receive special protection. Like many other nations of the industrialized world, Italy has recognized the need to act now to ensure the future of its natural and man-made wonders.

Still majestic despite the ravages of time, the amphitheater in Rome called the Colosseum is one of the largest relics of the Roman Empire. It held 40,000 to 50,000 spectators, who witnessed religious rites, chariot races, gladiatorial combats, and executions.

3

Italy Before the Renaissance

The earliest inhabitants to flourish in the Italian peninsula were the Etruscans, a seafaring people of uncertain origin. The ancient Greek historian Herodotus reported that the Etruscans launched their ships for Italy in the 12th century B.C. from Lydia in what is now the country of Turkey. But modern archaeologists and historians speculate that the Etruscans were either native to Italy or perhaps migrated there from Greece or Egypt. Some of this mystery is due to the Etruscan language, which used the Greek alphabet but has remained untranslatable.

Upon reaching Italy, the Etruscans built cities in the west-central part of the peninsula. Their state, which was called Etruria, controlled all the land along the west coast between present-day Rome and Florence. From the 8th to the 6th century B.C., Etruria dominated the area, including Rome; and even in decline, the Etruscan culture flourished into the 1st century B.C., when it ultimately became part of the dominant Roman Republic.

The Etruscans built towns surrounded by stone walls—these self-governed towns were the first version of the Italian city-state.

This fresco, or wall painting, is located at a site called the Tomb of the Leopard in Tarquinia, north of Rome. Tarquinia was a center of the Etruscan culture that dominated the central Italian peninsula before the rise of the Romans. Etruscan artifacts are noted for their grace and craftsmanship; the Etruscan language, however, has defied scholars' attempts to decipher it.

To avoid pirate attacks, the Etruscans built their settlements at least 5 to 10 miles (8 to 16 kilometers) from the coasts. Pirates themselves at first, they eventually abandoned piracy, building numerous harbors and establishing sea trade with Phoenicia, Greece, and Egypt. Apart from trade, the Etruscans excelled in architecture, metalworking, agriculture and irrigation, sculpture, pottery, and painting.

Etruscan society consisted of three classes: slaves, a large middle class of artisans and merchants, and an aristocracy. Among the aristocracy, the family was the most important social unit, and a woman's place in society was more highly respected than it was in ancient Greece. Women attended social and public gatherings; Greek writers criticized this aspect of Etruscan culture, but it was adopted by the later Romans.

Beginning in the 8th century B.C., another power, Greece, colonized the southern part of Italy. This region, called Magna Graecia (Greater Greece), was governed by dictators who ruled from Sicily. Syracuse, in Sicily, became the largest city in the Greek colony, with a population of 500,000. One of the most notorious of the Syracusan tyrants was Dionysius the Elder, who lived from about 430 to 367 B.C. He so feared assassins that he slept in a different room each night, and he imprisoned his enemies in a large cave that was called Dionysius's Ear because he eavesdropped upon them there. Today, visitors can still hear the eerie echoes produced by the cave's high walls.

In the 5th century B.C., the Greeks from Magna Graecia and a people called the Gauls from what is now France invaded Etruria, marking the beginning of the decline of Etruscan power. In the 4th century B.C., Etruria came under Roman rule, but it was not until 80 B.C., when a Roman general devastated the last Etruscan cities and gave Roman citizenship to all Italy, that Etruria finally disappeared.

The Kingdom of Rome

About the time Etruria was at its most powerful, Rome was a new city at the beginning of its history. The founding of Rome has been attributed to the legendary brothers Romulus and Remus, sons of the war god Mars, who were abandoned at birth and raised by a wolf. (The image of the wolf, which became the symbol of Rome, was originally borrowed from an Etruscan figure called the Lupa Capitolina.) Although it is impossible to separate long-ago history entirely from myth, it seems clear that the town of Rome was founded sometime in the 8th century B.C. on a hill called the Palatine along the Tiber River. Eventually, Rome would grow to cover six other nearby hills as well.

The Roman people are believed to have come from a merging of two similar west-central Italian cultures, the Latins and the Sabines. Latin kings ruled Rome until the Etruscans invaded the city in the late 7th century B.C. The final four monarchs of the Roman kingdom were Etruscans, but their rule ended in 510 B.C., when the Senate of Rome exiled the last king and established a new republic.

The Republic of Rome

The republic was governed by a Senate composed of patricians and plebeians (the aristocracy and the ordinary citizens, respectively). Rome also had a slave or peasant class, which had no voice in the government. Two consuls served as the chief magistrates of Rome, and there were times when one consul was permitted temporary dictator status.

The republic lasted approximately 500 years. During this period, its control was extended over most of the Mediterranean world. From the 5th to the 3rd century B.C., the Romans concentrated on

A circus scene with lions and a fallen gladiator or slave is depicted on a Roman decorative panel, now in the Museo Nazionale delle Terme, Rome.

conquering the Italian peninsula. They suffered occasional setbacks, such as an attack by the Gauls from across the Alps in 390 B.C. that resulted in the sacking of Rome and a rebellion supported by Greece south of Rome. Nevertheless, Rome had assimilated all of what is now central and southern Italy by 250 B.C.

The Romans next went after enemies across the waters. Between 264 and 146 B.C., Rome started and won three large wars against the prosperous city of Carthage, which lay across the Mediterranean Sea on the coast of North Africa. Carthage was the capital of the Phoenician, or Punic, empire, and these wars are called the Punic Wars. In the First Punic War, Rome and Carthage fought over the island of Sicily. Rome won, but only after a devastating 23-year struggle that killed about 20 percent of its population. The Second Punic War was waged for control of the Iberian peninsula (present-day Spain and Portugal), but it turned into a war for survival when a Carthaginian general named Hannibal crossed the Alps on elephants and terrorized Italy for eight years. He was defeated by the Roman general Scipio Africanus at the Battle of Zama in 201 B.C., and the war ended with Rome driving the Carthaginians from Spain and forcing Carthage to become an ally of Rome. This arrangement lasted for 50 years, until Rome, fearing that Carthage might regain its economic and military power, ordered the city destroyed. The Carthaginians resisted their fate for three years in the Third Punic War, but Rome eventually razed the city to the ground. During this period, meanwhile, Rome fought three wars in the eastern Mediterranean and conquered the entire Greek world.

Rome had become the only real power in the Mediterranean, but the republic did not survive to enjoy its good fortune. Centuries of war had caused a permanent economic crisis at home on the Italian peninsula, and Rome was filled with angry, dangerous mobs of unemployed peasants. Battles over reform measures—such as land redistribution and Roman citizenship for the poor—led to a rift in the Senate and eventually to civil war.

The Roman general Scipio (left) meets the Carthaginian general Hannibal (right, with elephants) at Zama in this fanciful scene, in which gods and cherubs look on. The Battle of Zama ended the Second Punic War and established Roman control over Carthage.

Furthermore, slave uprisings were common between 134 and 71 B.C. These uprisings, called the Servile Wars, were inspired by leaders such as Spartacus, a runaway slave who had been trained as a gladiator to fight other slaves and wild animals for the amusement of the public. The uprisings were violently suppressed by two generals, Crassus and Pompey, who emerged from this period of civil anarchy as two members of a triumvirate (a three-member committee of rulers). Together, they put down the revolts in 71 B.C. by ordering the crucifixion of 6,000 slaves.

The third leader to join Crassus and Pompey in the triumvirate was Julius Caesar. He had already distinguished himself in the government, where he developed a more accurate system of dates that is still called the Julian calendar, and on the battlefield, by conquering Gaul (now France) and Great Britain. He proved an able leader despite reports that he was prone to comas, nightmares, and epileptic fits.

The rule of the triumvirate did not last long. Crassus died, and in 49 B.C. Caesar defeated Pompey in battle and entered Rome in triumph. He named himself dictator for life and started on a plan to reorganize Rome, but some of his former allies, unhappy that Rome was changing from a republic back into a monarchy, plotted against him. On the day called the Ides of March (March 15) in 44 B.C., Caesar was stabbed and killed in the Senate. But the republic could not be salvaged.

The Empire

During the years of anarchy that followed, Rome was torn between self-government and monarchy. In 27 B.C., Caesar's nephew, Octavian, emerged from the civil wars as the first emperor and took the name Augustus Caesar. Augustus was not a military man like his uncle. Instead, his successful rule was due to his administrative skill. His reign brought peace to Rome's vast empire, from the British Isles and Spain to North Africa and the Middle East.

The two centuries of peace that started with the rule of Augustus are known as the Pax Romana. During this period the Roman economy was stabilized; great public buildings such as the Roman

Julius Caesar conquered France and Britain for Rome and made himself dictator of the republic before being assassinated in 44 B.C. His nephew Octavian was to become the first Roman emperor.

Colosseum were built, as were paved roads that were still in use a thousand years later; and writers such as Virgil and Horace created the first great Roman literature. The city of Rome grew and flourished, until its population was about one million by the 2nd century A.D.

Augustus did more than start the Pax Romana, however. He also sowed the first seeds of the downfall of the empire, for he began the practice whereby each emperor chose his successor from among the members of his own family. For the following two centuries, Rome's rulers were of all sorts: Some were competent and sober, others were extravagant or even mentally ill. During the first century, the emperor Nero was particularly degenerate. He is said to have poisoned his mother, kicked to death his second wife, burned part of Rome to the ground, and killed the first Christians, including Peter and Paul.

The 2nd century A.D. is called the Golden Age of the Roman Empire because the emperors of this period were for the most part well-respected statesmen, soldiers, and thinkers. Rome's borders were extended into what is now part of Eastern Europe and the Middle East, and this expansion was accompanied by a great deal of construction throughout the empire—waterways called aqueducts were built in many areas, as were temples, theaters, and roads.

This prosperity continued until the 3rd century, when a succession of more than two dozen corrupt and brutal emperors weakened the empire through greedy, oppressive, or inept government. Economic collapse and invasions from the north and the east resulted in a ruined state dominated by rich landowners, with the large peasant class stripped of all rights and freedoms.

Constantine the Great was one of six rival emperors who vied for control of the empire in the early 4th century. He introduced religious tolerance to the empire with the Edict of Milan, a proclamation issued in 313 that allowed Romans to be Christians; in fact,

Made the official religion of the empire by Constantine in the 4th century, Christianity soon established itself throughout the Italian peninsula. The city of Ravenna is noted for its many examples of early Christian art, including these mosaics in the church of San Apollinaire Nuovo. The church was built by Theodoric, a Germanic chieftain who ruled Italy from 493 to 526.

he became a Christian and made Christianity the official state religion, which gave the church its first political influence. Constantine fought off the Goths (a Germanic tribe), the Franks (from present-day France and Germany), and the Persians. In 324 he emerged as the ruler of the whole empire. He rebuilt the ancient city of Byzantium in what is now Turkey, renamed it Constantinople (it is now called Istanbul), and made it the new capital of the Roman Empire. This proved to be the ruination of Rome, for the empire split into eastern and western halves, one ruled from Constantinople and the other from Rome. The eastern empire (called the Byzantine Empire) grew and prospered, but the western empire grew weak and fell prey to numerous attacks by Germanic tribes such as the Goths and the Vandals. In 476 the last western emperor was deposed by the Germanic chieftain Odoacer, and ancient Rome came to an end.

Invasions and the Church

Invasions from the east and north had occurred throughout Rome's history, but invading groups held lasting power in Italy after Rome's fall; as recently as the 19th century, other European nations were carving up the Italian land for themselves. After the last Roman emperor was deposed by German invaders, the eastern emperor in Constantinople turned to the Ostrogoths, another Germanic tribe, for help. They recaptured Italy for the empire and made Ravenna the new capital. Meanwhile, the Lombards, still another Germanic tribe—led by King Alboin—arrived in the Po River valley in 568; they established a capital at Pavia and ruled north and central Italy for the next 200 years.

Pope Leo III crowned the Frankish king Charlemagne first Holy Roman Emperor in 800. Charlemagne's empire, which included Italy, broke up after his death, and Italy entered a long period of invasions and disunity.

Another force that shaped Italy's history after the fall of the Roman Empire was Christianity. Christian communities had existed in Rome as early as the middle of the 1st century A.D. Beginning in the 4th century, self-sufficient religious communities called monasteries came into existence, partly as a response to the barbarian invasions. Within the walls of the secluded monastery, the devout Christian escaped unhappy worldly matters and carried on a quiet life dedicated to work and contemplation. St. Benedict of Nursia, a monk who founded the religious order the Benedictines, described the function of the monastery in a document called the Benedictine Rule written toward the middle of the 6th century. This document was used as the model for western monasticism for more than 500 years.

At the same time, the papacy was taking shape under the guidance of Pope Gregory I (590–604). Gregory was a monk at one of the monasteries he had built on his family's property in Sicily. He increased papal influence by sending monks to convert the Anglo-Saxons in England and by making the eastern bishops subordinate to Rome. As Christian communities gained power and influence in central Italy, he oversaw their development. By the time he died, Pope Gregory I had forged the papacy into a great religious and political power.

Because they were close to the Lombard territory, however, the new papal communities were under constant threat of invasion. When religious conflict spread from Constantinople and the Byzantine Empire to Italy, violence broke out in cities such as Rome, Naples, and Venice, and this allowed the Lombards to increase their influence in papal territory. Finally, Pope Stephen II (752–57) enlisted the aid of the Frankish tribes that were sweeping through the peninsula. The Franks conquered the Lombards and helped establish the Papal States. In return, Pope Leo III (795–816) crowned their king, Charlemagne, Emperor of the West in 800.

After Charlemagne died, his empire disintegrated under the onslaught of recurring invasions from abroad, particularly from the Saracens (Islamic warriors from North Africa and the Middle East). The Papal States once again needed protection, and in 962 the pope made another Germanic king, Otto I, the first Holy Roman Emperor in return for his help. The Holy Roman Empire lasted for nearly 850 years, but the Ottonians lost their power soon after Otto's death, and the northern half of Italy was taken over by merchants and landowners, who created the medieval city-states.

Called communes, these independent city-states began to spring up in the 11th and 12th centuries. Chief among them were Genoa, Pisa, Venice (which had been an independent republic since the 7th century), Florence, Siena, and Milan. They were able to maintain their independence because they were thriving, self-contained commercial centers with no need for support from a central government. The city-states were organized and governed by guilds of merchants, professionals, and craftsmen.

Sicily and Southern Italy

Sicily passed through many hands in these centuries. The Saracens established themselves on the island in the 9th century. In the 11th century the island was taken over by the Normans (a people from the western coast of France who were conquering Great Britain at the same time). The Normans also conquered territory in southern Italy; the region they controlled there was known for a time as the Kingdom of Naples.

As a result of marriage between royal families, Sicily and the Kingdom of Naples became part of the territory of the Holy Roman Emperor Frederick II (1194–1250). Frederick challenged the power of the pope, sought independence from papal influence for his kingdoms, and fostered a burst of cultural development. A patron of poets and philosophers, he spoke six languages, financed the translation of texts from Arabic and Greek, wrote the first modern book of natural history (on falconry), and founded the University of Naples in 1224.

Frederick's reign was filled with political conflicts and land disputes. He was often in trouble with the church; two popes excommunicated him and one tried to have him deposed. One excommunication was for refusing to lead a crusade against the Moslems in Jerusalem (eventually he agreed, captured Jerusalem, and crowned himself king), and the other was for the general charge of heresy. Frederick was still fighting the popes on his deathbed in 1250. He had tried to claim northern Italy as part of his kingdom and had waged a series of unsuccessful battles with the Papal States and the independent city-states. People who sided with him were called the Ghibellines (an Italian version of a German battle cry), and those who opposed him were known as the Guelfs. The conflict between these two factions raged on after Frederick's death and became so widespread and destructive that in 1309 the pope moved the seat of the papacy from Rome to Avignon in southern France.

After Frederick's death Sicily fell on hard times. By 1266, the pope had placed a French king named Charles I in power. His tyranny inspired a rebellion in Sicily; the people there made Peter III of Aragon (a province in Spain) their king, while Charles I continued to rule the Kingdom of Naples. Not until the 15th century would the island and the mainland be reunited in one kingdom.

Hilltop fortresses such as this castle in the city of Campo Tures, northwest of Bolzano, near the border of Austria, were the nuclei of the Italian city-states that developed after the fall of the empire. These independent, self-sufficient communities were the birthplace of the Renaissance.

4

The Renaissance
and After

Italy was fragmented into a patchwork of rival kingdoms and communes after the fall of the empire, and it was not to be unified until the 19th century. Separate political regions developed around some of the major communes, and these city-states hosted a surge of cultural and economic activity that eventually influenced all of Western Europe. But their lack of political unity left them prey to the larger, more aggressive powers of France, Spain, and Austria.

By about the 13th century, the Italian city-states run by merchants and traders created an elite urban consumer environment, with new products, fashions, and ideas. In order to celebrate their growing economic power and relative social stability, the middle and upper classes looked back to the golden ages of the Greek and Roman empires, which they saw as ideal civilizations, and tried to re-create the arts and customs of these bygone eras. The result was the Renaissance (French for "rebirth"), a period of rapid cultural and scientific growth that lasted roughly from the 14th through the 17th centuries. Although the Renaissance affected all of Europe, it started in Italy.

Renaissance culture valued learning above all; Italian universities attracted students from all over Europe. These pupils studied Latin, Roman history and archaeology, and classical mythology and literature. A philosophy called humanism developed—it emphasized individual interests and a view of life based on worldly experience and learning rather than on religious dogma.

The great artistic and intellectual flowering of the Renaissance is probably best represented in the life and work of Leonardo da Vinci (1452–1519). In an application for a position at the court of Milan, Leonardo outlined the range of his skills: As an engineer, he could build bridges and moats; he also promised to construct war machines such as guns that fired stones and armored chariots. In

Plague swept through Europe in the 14th century, killing as much as one-third of the continent's population. The Black Death, as it was called, inspired artworks such as this image of Death seated on a skeleton horse, striking down kings and queens, bishops and troubadors, along with peasants and townspeople.

peacetime, he could build palaces, churches, and aqueducts. He could create sculptures in marble, bronze, and clay. Finally, the man who was perhaps the greatest of the world's painters added casually, "Also I can do in painting whatever may be done." Learned and skilled men like Leonardo gave rise to the modern expression "Renaissance man," which is used to describe someone of many abilities and interests.

Plagues and Politics

Along with the rest of Europe, Italy was ravaged in the middle of the 14th century by the Black Death, a combination of plagues—chiefly the bubonic plague—that were carried to Genoa by Italian merchants returning from trading voyages to the Middle East. The plagues spread rapidly and people succumbed in great numbers, for Europeans had never been exposed to these diseases and had developed no immunity to them. It is estimated that between 1347 and 1351 the Black Death killed a third of the population of Europe.

Politically, the Italian city-states of the Renaissance were rather anarchic. Each was run by its richest families. Smaller cities consolidated with larger ones, until four northern city-states developed into formidable powers: Milan, Florence, Venice, and the Papal States. They vied for power, made and broke alliances with each other and with other European leaders, and were generally chaotic. Their inhabitants prospered, however, becoming the best-educated, most privileged people in Europe.

Although Florence had a republican form of government, it was actually run by a small group of wealthy families. Chief among these was a powerful banking family called the Medicis. One Medici, called Lorenzo the Magnificent (1449–92), was a brilliant statesman, diplomat, and patron who ruled Florence during its greatest period, the late 15th century. The Medicis funded hospitals, subsidized churches and monasteries, and revitalized universities. They founded the first public library in Europe, stocked not only with books imported from other collections but also with manuscripts from the Far East and new translations of important Greek and Arabic texts. The Medicis invited painters and sculptors to study their collection of classical antiquities, which included everything from gems to statuary. In return, dozens of artists offered their services to the great family, designing and decorating their grand villas and also adorning Florence with fine artworks. It is no wonder that the city and its most illustrious patrons became the model and envy of all Italy. The example of the Medicis was followed by the Este family in Ferrara, the Montefeltros in Urbino, the Gonzagas in Mantua, and especially the Sforzas in Milan.

The Sforzas were the most important family in Renaissance Milan. Like the Medicis, they were patrons of art and education, and they helped make Milan one of the richest states in Italy. In particular, they supported Leonardo da Vinci. But the Sforzas, like many ruling families of the era, were prone to excesses and to a craving for more power and property. They were involved in many intrigues among the powerful people of the time, including the strong and ruthless Borgia family of the Papal States. In the 16th century, as Milan began to go bankrupt, the Sforzas fled to Austria, while France—with the help of the Papal States under the command of Cesare Borgia—took control of the city.

Another important Renaissance leader was Pope Julius II, who held the papacy for 10 years, from 1503 to 1513. He was called the Warrior Pope, for he united with Venice and Spain to drive the French invaders out of Italy. He also set about restoring Rome to its former eminence. Rome had long been venerated as a holy city. For Christian pilgrims, all roads led to Rome; within its walls lay the tombs of saints Peter and Paul and, since 1462, the sacred skull of St. Andrew (Peter's brother). In the years before Julius rose to the papacy, however, the city had fallen into neglect, a condition the new pope found intolerable. Italy boasted the finest artists in the world. He enlisted their help in rebuilding and enriching Rome,

Lorenzo de' Medici, known as Lorenzo the Magnificent, ruled Florence during the late 15th century, when the city reached the heights of artistic and political importance.

which was called the Eternal City. Some of the most celebrated talents of the time—in fact, of all time—answered the call. Michelangelo painted the ceiling of the Sistine Chapel in the Vatican, the immense papal headquarters. Raphael, another noted painter, decorated the papal chambers. An architect named Bramante drew up plans for a new basilica to be built on the site of St. Peter's tomb. Architects and engineers set about widening bridges and redirecting roads to accommodate the swelling throngs that began to pour into the Holy City.

Reformation and Counter-Reformation

While Rome was reaping the artistic harvest of the Renaissance, the humanist philosophy of the period was clashing with the church, which had become exceptionally powerful and influential. Conflicts between free thought and official church dogma resulted in battles of ideas over the next few centuries, beginning with the Protestant Reformation and the Catholic Counter-Reformation and continuing with the development of science and the Enlightenment.

The immense basilica of St. Peter was planned and commissioned by Julius II, who became pope in 1503 and called on Italy's architects and artists to embellish Rome and the papacy. In addition to St. Peter's, the works carried out under Julius's guidance include Michelangelo's Sistine Chapel ceiling.

The Protestant Reformation was started by Martin Luther, a theologian from Wittenberg, Germany, who had made a pilgrimage to Rome. Instead of the spiritual enlightenment he expected, he discovered moral apathy and degeneration within the Roman Catholic church. The clergy seemed to be more interested in accumulating wealth and power than in serving God. Luther left with a sense of outrage. When he returned to Germany he issued a radical statement, demanding that the German church break all ties with Rome. In 1517 he committed one of the most celebrated acts in history: He posted on the door of a Wittenberg church a list of statements that spelled out his objections to the Roman Catholic church. Luther's spirit of reform swept through Europe, and the Protestant Reformation was born.

In the next decade, Spain invaded and conquered Rome. Under Spanish influence, the papacy established an ultra-Catholic reaction to Martin Luther; this was called the Counter-Reformation. It was inspired by an order of Spanish Catholics called the Society of Jesus, or Jesuits, who opposed Luther's reforms and advocated strict religious discipline and unwavering obedience to the pope. One product of the Counter-Reformation was the Inquisition, a tribunal run by the Jesuits that sought to destroy Protestantism and to punish heretics, including Jews and intellectuals who questioned the authority of the church.

Protestant ideas were not the church's only challenge in the 16th century. Another threat came from thinkers who labored outside the monastery or the chapel. For a long time, it had been possible to include the theories of ancient scientists such as Euclid, Ptolemy, and Aristotle in standard church doctrine. But the growth in learning had spawned new thinkers and new ideas. Scientists at universities in Bologna, Padua, and Pisa arrived at conclusions—based on their own observations of nature or on laboratory experiments—that disputed the church's view of heaven and earth.

When Martin Luther nailed his statements to the door of a church in Wittenberg in 1517, he not only launched Protestantism but also set in motion religious conflicts that would challenge the power of the papacy.

For centuries, for example, no one had challenged the Bible-based notion that the universe was a closed system created by God with the earth at the center. But Copernicus (1473–1543), a Polish astronomer trained at Italian universities, proposed a different picture of the heavens. He suggested that all the planets, including Earth, revolved around the sun. Copernicus knew how upsetting his theory would prove, so he made sure it was not published until after he died. He thus eluded the wrath of the Inquisition. Instead this wrath fell on his supporters. One of these supporters was Galileo Galilei (1564–1642), a brilliant Italian scientist who had already created some controversy by demonstrating—from atop a leaning tower in the city of Pisa—that objects of different weight dropped from the same height reach the ground at the same instant. Churchmen felt that there was no need for experiments such as this; all that humankind needed to know about natural laws was contained in the Bible. But in 1623, Galileo published his own theory of the universe, which agreed with Copernicus's findings: The earth was not the center of creation.

The church retaliated quickly. Galileo was brought before the Inquisition and was forced to renounce his writings—that is, to state that he had been in error. But it was too late. His ideas were already circulating among other thinkers, and they helped form the foundation of modern science.

Science and education continued to grow throughout Europe, and gradually the church's influence dwindled. By the 18th century, a new era of thought had dawned. It was called the Enlightenment, and the thinkers and writers of this movement left church teachings behind and tried to reach a better understanding of nature, society, and humankind through logic. Two Italians in particular contributed to the Enlightenment. One was Cesare Beccaria, whose *Essays on Crime and Punishment* led to broad reforms in the treatment of prisoners and criminals. The other was Giambat-

A fresco by Luigi Sabatelli shows Galileo demonstrating his telescope to the Senate of Venice. Although the Inquisition condemned Galileo's work as heretical, his theories survived as one of the cornerstones of modern science.

tista Vico, who developed a theory of historical cycles that said that all cultures must experience periods of growth, decay, and rebirth; this theory became widely accepted in the 19th and 20th centuries.

The European Power Struggle

The European nations entered a long period of strife at the end of the Middle Ages, and Italy was part of their power struggle. France and Spain in particular continued to play tug-of-war over Italian territory throughout the Renaissance and after.

In 1527, Charles V of Spain sent troops to invade Rome. They laid waste its buildings, plundered homes and churches, flung orphans and the bedridden into the Tiber River, and kidnapped and assaulted women. Pope Clement VII, whose alliance with Francis I of France had provoked Spain to attack, fled the Vatican and hid in the Castel Sant'Angelo, the ancient tomb of the Roman emperor Hadrian.

56

Napoléon Bonaparte, the Corsican soldier who became emperor of France and ruler of most of Europe, commanded France's army in Italy in 1796 and 1797.

Meanwhile, battles were being waged on other fronts. When Louis XIV of France seized the Spanish crown, those two countries ended their war. In 1713, Italy was carved up and divided among the European powers by a treaty called the Peace of Utrecht. Austria received most of Spain's possessions in Italy, and the duke of Savoy (a small principality on the border between France and Italy) got the island of Sicily, although he soon swapped it for the Kingdom of Sardinia. In 1734, Spain grabbed Naples and Sicily.

In the late 18th century a new figure emerged to dominate European politics. This was Napoléon Bonaparte (1769–1821), a native of Corsica, an island located between Italy and France. As a young man, he joined the French army, and in 1796, after the French Revolution, he was named commander of France's army in Italy. In 1804, he became the emperor of France, summoning Pope Pius VII from Rome to preside at his coronation in Paris. By 1810, Napoléon ruled most of Europe.

Napoléon brought many innovations to Italy. He established nationwide laws, installed efficient local governments, and supervised the construction of roads and bridges. For a time, the Kingdom of Italy was ruled by Napoléon's stepson Eugene, and the centralized state that the emperor created in Italy—although it was short-lived—inspired the first stirrings of nationalism in the country.

Napoléon lost his empire in 1814, when the British defeated him at Waterloo. In 1815, the European powers met at the Congress of Vienna to determine the fate of the Continent. Italy fared badly. Once again it was carved up and apportioned to various nations. Most of northern Italy went to Austria; Piedmont (a region near the western border with France) was given to Savoy; the old Kingdom of Naples went to the Spanish king; and the Papal States were handed back to the pope.

This carving-up fed the newborn feelings of nationalism that were shared by a growing number of Italians. Secret pro-independence societies sprang up. One was called the Carbonari (Charcoal Burners), another the Spillo Nero (Black Pin). These and other groups wanted a united Italy and an end to outside interference. The Italians rebelled against their foreign oppressors several times between 1820 and 1832, but the rebellions failed. One problem with the nationalist movement was that its members disagreed in their goals for the country. Some of them wanted to establish an Italian republic, some wanted to make the pope president, and still others were monarchists, who wanted Italy united under a king. The monarchists were the most powerful group among the nationalists. They led the movement that resulted in the birth of modern Italy.

During the rule of Benito Mussolini, the children of the mountain village Capalbio posed beneath a portrait of the dictator and his motto, which reads, Believe, Obey, Fight.

5

Modern History

The movement to achieve a unified Italy was called the Risorgimento. It succeeded because of the efforts of three very different men: Giuseppe Mazzini (1805–72), Count Camillo Benso di Cavour (1810–61), and Giuseppe Garibaldi (1807–82).

Mazzini was a former member of the Carbonari. Between 1831 and 1857, he became the leader of the republican movement and organized rebellions in Piedmont, the Papal States, and southern Italy. He was exiled to France, where he formed the Young Italy Association. Its goal was to create a unified Italy, "one free, independent nation." The association, which would not allow anyone over the age of 40 to join, had 60,000 members in 1833.

Cavour was an aristocrat who by birthright should have risen to the top of the Piedmontese court. Instead, he came to believe in social change. After traveling throughout Europe, observing social conditions, Cavour returned to Piedmont in 1847 and became publisher of the newspaper *Il Risorgimento*. In 1852, he assumed the post of prime minister of Piedmont.

Cavour knew he needed outside help to chase the Austrians out of northern Italy. In 1858, he met secretly with Emperor Louis-

Napoléon of France, and they drew up a plan. King Victor Emmanuel II of Piedmont was to become king of the lands between the Alps and the Adriatic Sea. In return, France would receive Nice and Savoy once the combined forces of France and Piedmont reclaimed them from Austria. To provoke Austria into war, Cavour increased the size of the army in Sardinia. Austria attacked, and Louis-Napoléon and his French troops rallied to the Italian side.

In July 1858, Austria was driven out of Italy, and Louis-Napoléon signed the Peace of Villafranca with Emperor Francis Joseph I of Austria. The terms of the agreement came as a crushing disappointment to the Italians: Austria would retain much of northern Italy, and Italy was to become a confederation under the pope. In outrage, Cavour resigned. But he did not sit idle for long. He returned to office in 1860 and demanded that the regions in question be allowed to settle their own destiny by vote. Savoy and Nice chose to become part of France, but several other disputed regions joined the Italian state.

Meanwhile, the third architect of Italian independence had launched a campaign of his own. As a member of the Young Italy Association, Garibaldi had traveled to South America in 1835 and fought in the Uruguayan civil war. He lived for a short time on Staten Island, New York, before returning to Italy in 1851 as a firm ally of Victor Emmanuel II. In 1860, heading his own military force of about 1,000 men, known as The Thousand or the Red Shirts, he drove a much larger force of Spanish troops from Sicily and Naples. With the elimination of foreign forces from the south and much of the north, the stage was set for the unification of Italy. In 1861, the Kingdom of Italy was proclaimed, with Victor Emmanuel II as its first king. The kingdom was a constitutional monarchy whose first parliament met in 1861.

From the beginning, Italy's principal goal was to annex—that is, to obtain control of—the handful of Italian territories that remained

On May 15, 1860, a thousand volunteer soldiers called Red Shirts routed a much larger force at the Battle of Calatafimi, setting the stage for the patriot Garibaldi's victory in Sicily. This painting of the battle is housed in the Risorgimento Museum in Milan.

outside the kingdom. Five years later, Italy reclaimed its northern territories from Austria in the settlement that followed the Austro-Prussian War. In 1870, the Italian government seized Rome, which had remained in the hands of the pope. By the following year, Italy was completely united and Rome was its capital.

But a lasting union was hard to maintain in a country made up of regions that had very different histories, political goals, and dialects. In fact, many Italians resisted the federal government and refused to put national interests above regional ones. Confusion resulted, and the economy suffered. In 1861, the national debt was 2,450 million lire; 4 years later, it had doubled.

Other problems came from outside. Railroads now traversed much of Europe, which meant that other, more industrialized nations could send their mass-produced goods to Italy, and this hurt

Italian producers whose goods were made less efficiently and therefore more expensively. The economic picture grew so bleak that many Italians emigrated. In the decade that preceded World War I, 16 out of every 1,000 Italians left their homeland to seek their fortunes elsewhere, often in the United States.

The Rise of Imperialism

The people's dissatisfaction over economic troubles lent strength to political parties that opposed the government. In 1876, the left wing came into power behind Agostino Depretis, who headed a corrupt government. Another prime minister of this period was Francesco Crispi. He held the office twice and used extreme force to suppress violent outbreaks in Sicily. Crispi also nurtured grandiose dreams. England and France had become colonial powers, with holdings and territories scattered around the globe, and he wanted the same for Italy. In 1889, Italy established control over Somaliland, a region on the east coast of Africa, and in 1890 it did the same in another East African land that it named Eritrea. Crispi overreached himself,

Italian soldiers celebrate the capture of the Ethiopian provincial capital of Mekele in 1935. Italy succeeded in seizing control of Ethiopia on this, its second, attempt but lost its claim to the African country after World War II.

however, and lost a war against Abyssinia (present-day Ethiopia). This defeat brought down his government.

The next prime minister was Giuseppe Giulotti, who held the top office on and off from 1901 to 1914 and again in 1920. He initiated social reforms, raised production and wages, and balanced the budget. During his administration, dreams of Italian expansion were revived. A party called the Nationalists urged the annexation of Libya, a territory in northern Africa that had once belonged to the Roman Empire. Italy invaded Libya in October 1911 and retained control of that country until World War II; many Italians settled there.

Throughout this period, Italy was plagued by social and revolutionary ferment. These troubles were at the worst at the end of the 19th century, when riots broke out in Rome and in the south. The rioting spread to other regions, and the military restored order—but not very successfully. In 1892 the Italian Socialist party (PSI) was founded in Genoa and adopted a Marxist program of opposition to the government. King Umberto I was assassinated in 1900. In 1908, Prime Minister Giulotti violently quelled strikes by agricultural workers in Parma and Ferrara.

Meanwhile, Europe was moving toward World War I, which broke out in 1914. Italy, part of the Triple Alliance—along with Germany and Austria—proclaimed its neutrality, but many Italians disagreed with that decision. One group, called the Interventionists, wanted Italy to enter the war on the side of the Allies (France and England). An opposing group, the Neutralists, favored the official policy of noninvolvement. The other European nations pressed Italy to enter the war and competed for its support. Finally, on April 26, 1915, Italy joined the Allies and declared war on its old foe, Austria. According to a provision in the Treaty of London, if the Allies won, Italy would reclaim lands south of the Alps that were still in Austrian hands.

It was one thing to sign a treaty, however, and another to fight a war, especially one of this magnitude—the first war fought with airplanes, tanks, and nerve gas. Italy's army was by no means prepared for this. It was short on equipment and weak in organization. In addition, the government was hard-pressed to stir up any enthusiasm for the war among the populace. Yet Italy contributed to the Allied victory. On October 24, 1918, its army pinned the Germans and the Austrians behind their own frontiers. In November, Italy and Austria signed a peace agreement.

On June 28, 1919, the Treaty of Versailles spelled out the terms of peace in Europe. Italy left the treaty table fattened but still hungry. The Italian negotiators had won the northeastern provinces of Trentino, Alto Adige, and Trieste, but they had failed to get two other provinces, Dalmatia and Fiume.

Mussolini and Fascism

After the war, Italy was a shambles. Its debt was high, and its currency—the lira—steadily lost value. Prices soared, working conditions were abominable, and unemployment grew as war veterans swelled the work force. There were massive strikes, often put down by armed security police called *fasci* who were hired by factory owners and landowners.

The word *fasci* comes from a Latin term that described the bundles of rods Roman officers tied to their axes. These rods became a symbol of the strong-armed squads used by the Roman state to uphold its authority, and the term *fascism* was used to describe authority maintained by force. In the years after World War I, a network of groups came together under the banner of fascism. These groups included Interventionists, Nationalists, monarchists, anarchists, revolutionary socialists, and others who differed in their political theories but shared one belief: They favored violence as a means of achieving their ends. All they lacked was a leader. Eventually, they found one.

(continued on page 73)

SCENES OF
ITALY

Overleaf: *Located on the rugged west coast, south of Naples and Mt. Vesuvius, the small town of Positano is a resort popular with Italians and tourists alike.*

Fishermen from Pozzuoli, near Naples, sort their day's catch. Italy's fishing industry harvests great numbers of anchovies and sardines from Mediterranean waters. Bluefin tuna and squid are other important food catches, and sponges are gathered for export.

The emperors Augustus and Tiberius built splendid vacation villas on the island of Capri, south of Naples. Then as now, Capri lured visitors with its breathtaking views of the surrounding sea. Its most-noted tourist attraction is a partially flooded cave called the Blue Grotto.

Near Ninfa, a small town in the northern lake region, roses and pines make a lush garden in the ruins of a house (or barn). The Italian countryside has been much admired for its ability to evoke a sense of the past.

In the city of Vicenza, west of Venice, revelers in traditional costumes prepare for the Marostica, a festival in which participants enact the roles of chess pieces on the large checkered square in the background.

A gondolier steers his passengers past the sights of Venice.

Students, shoppers, tourists, and passersby of all sorts enjoy spending a quarter of an hour—or a whole afternoon—in Rome's sidewalk cafés. This one is located in a square called the Piazza Navona.

70

A 1604 painting by Caravaggio shows the body of Christ being taken down from the Cross. Caravaggio's dramatic use of light and shadow greatly influenced later European painters; this work, in fact, was copied by the 17th-century Flemish painter Peter Paul Rubens and the 19th-century French painter Paul Cézanne.

As artist, scientist, and engineer, Leonardo da Vinci epitomized the Renaissance spirit of creativity and inquiry. He is thought to have painted this portrait of Ginevra dei Benci, a Florentine noblewoman, around 1474 (a clue to her identity is found in juniper leaves sketched in the background and on the back of the canvas; Ginevra is Italian for "juniper").

The Vatican Museum houses one of the world's great art collections. One of its treasures is this 14th-century altarpiece painted by Giotto that depicts St. Peter enthroned between angels and saints.

The city of Rome owes much of its beauty and grandeur to Pope Julius II, who wanted it to be a showplace of Christian art. Michelangelo placed his marble statue of Moses over Julius's tomb; like Moses, Julius was honored as a guardian of humanity's spiritual welfare.

(continued from page 64)

Benito Amilcare Andrea Mussolini (1883 – 1945) was the son of a blacksmith who was a social activist. As a boy, Benito was a ruffian, prone to hurling stones at churchgoers and starting knife fights with classmates. His belligerence became the foundation of his political convictions. "Violence is not a system, not an aestheticism, and even less a sport," he wrote. "It is a hard necessity." As editor of the Socialist newspaper *Avanti!*, Mussolini found the ideal forum for voicing his rage against the church and the state and for practicing his considerable abilities as a rabble-rouser. But when he used the pages of *Avanti!* to support Italy's entrance into World War I, Mussolini was forced to resign the editorship. He then started his own newspaper, *Il Popolo d'Italia*.

Mussolini served in World War I, then became an organizer of fascism. He turned the offices of *Il Popolo d'Italia* into the head-quarters of his Fascio di Combattenti, or Battle Band, a gang of 150 men determined to eliminate its left-wing opponents. The Italian government wanted to suppress the political left and so tolerated Mussolini's thugs. In May 1921, Mussolini and 34 of his followers won seats in the Italian parliament, and in November the Fascist party was officially formed. In August 1922, Fascisti (members of Mussolini's party) raided the offices of *Avanti!*, now the opposition newspaper, and burned them to the ground. Mussolini and party members in Naples, now calling themselves the Black Shirts, pre-pared to march on Rome on October 28. A few days before then, the prime minister and the entire cabinet resigned and King Victor Emmanuel III submitted to pressure and let Mussolini take charge. The blacksmith's son replaced the cabinet with his own henchmen and proclaimed himself dictator, Il Duce.

This was just the beginning. Mussolini established the Corporate State, a central body that was to oversee every detail of Italian life. He scored remarkable successes. He increased employment, au-thorized huge building projects, overhauled mass transportation,

Dictator Mussolini addresses a crowd in the city of Treviso, near Venice. Mussolini admired the methods and philosophy of Germany's Adolf Hitler, with whom he signed an alliance that drew Italy into World War II.

and installed a program of social benefits. His government subsidized farmers and thus freed Italy of its dependence on imported wheat. In 1929 he devised the Lateran Treaty, which reduced the power of the church, a long-time enemy of the state. In return, however, Roman Catholicism was named the state religion. Mussolini also imprisoned many of his political enemies.

Mussolini was now determined to create an empire. He increased the nation's military strength, but the fighting force was still weak, so he had to choose his victims carefully. In October 1935, Italy attacked Ethiopia in order to avenge Italy's defeat there decades earlier. Poor as Mussolini's army was, Ethiopia's was poorer still. Its soldiers were barefoot and shouldered makeshift weapons. Italy won a quick victory, and Victor Emmanuel III was proclaimed emperor of Ethiopia. Mussolini's next foray was in

1936, when civil war broke out in Spain. Seventy thousand Italian troops came to the aid of Generalissimo Francisco Franco, the Spanish dictator who had rebelled against his government. Another supporter of Franco was Adolf Hitler, who headed Germany's Nazi party. In October 1936, Mussolini and Hitler formed an alliance called the Rome-Berlin Axis.

The two dictators had a strange relationship. They had met in 1934 in Venice. At that time, Mussolini, who had many admirers throughout the world, was the dominant figure, and Hitler played the role of protégé. Within a few years, their positions were reversed. Hitler had built Germany into a powerful economic and military force, and Mussolini spoke fervently of "Prussianizing" Italy—that is, of making it into a copy of Hitler's Germany. In September 1939, when Hitler invaded Poland and started World War II, Mussolini announced that Italy would remain neutral. But the two dictators conferred in March 1940, and Hitler convinced Mussolini that the fates of fascism and nazism were intertwined. On June 10 of that year, therefore, Mussolini declared war on France and England.

Italy found itself overmatched. Ill equipped, small in number, undisciplined, and short on spirit, the Italian army suffered heavy losses. When the troops crossed the snow-covered mountains into France, they were clothed in threadbare uniforms and cardboard shoes, and frostbite claimed as many lives as did enemy weapons. An attack on Greece in October 1940 was fiercely resisted, and the Italians retreated in humiliation. The following year, they lost Eritrea, Somalia, and Ethiopia to the English. In the winter of 1942–43, 100,000 Italians—half the armed force—died on the Russian front.

In July 1943, the Allies landed in Sicily. Joined by a growing number of partisans (Italians who opposed Mussolini's Fascist rule), they pushed northward, beginning a year and a half of fighting on Italian soil. At the same time, the Italian parliament

Allied troops, many of them from the United States, fought fiercely to drive German oc-
cupying forces out of Italy. Many towns and much of the countryside were devastated
by the combat. Here, the people of a village called Bojano return to their homes in 1943,
after an Allied victory.

voted Mussolini out of office and he was arrested. In October 1943, Italy formally renounced its connection with Germany and joined the Allies. By that time, 60,000 Italian partisans had perished fighting against nazism and fascism.

Mussolini had been hiding in the Alps with his mistress, Clara Petacci. On April 27, 1945, he was captured when he attempted to escape. He and Petacci were assassinated, and their corpses were put on public display in the Piazzale Loreto in Milan.

Once Again, a Republic

As at the end of World War I, postwar Italy was a mess. Battles had left the country in ruins, and the obligation to pay wartime reparations to the Soviet Union, Greece, Yugoslavia, Albania, and Ethiopia plunged Italy still more deeply into debt. Yet the country has made a phenomenal recovery in the decades since the war.

Financial aid came from the United States and other countries. In 1957, Italy joined West Germany, France, Holland, Belgium, and Luxembourg in establishing the European Economic Community (now called the European Union), which integrated the economies of member countries. From 1959 to the mid-1960s, Italy enjoyed the biggest economic boom of any nation that had fought in World War II with the exception of Japan.

In 1946, the monarchy was ended by popular vote and Italy proclaimed itself a republic. A new constitution came into effect the following year. Since that time, Italy's political history has been one of almost constant change.

The government of Italy is often described as sitting on the brink of disaster because its prime ministers change with such frequency; few of them remain in office for more than a year or two. And rarely does a single party control the government. Instead, most governments are coalitions—that is, temporary arrangements among two or more parties, none of which can win enough votes to assume full

The slain bodies of Benito Mussolini and Clara Petacci, his mistress, are hanged by their heels in a square in Milan in 1945. The dictator was captured while trying to flee the country.

After the radical Red Brigades kidnapped and murdered former prime minister Aldo Moro in 1978, the place where his body was discovered in a car became a shrine for mourners such as this woman, who touches Moro's portrait.

control. The number of coalition governments that the country has seen—more than 50 in the same number of years—adds to the appearance of instability. But Italian political life is more consistent and predictable than many outsiders realize. For example, Italians choose their political leaders from a small pool of people who are similar in many ways.

The 1960s was a turbulent decade in Italy, as it was in the rest of Western Europe and the United States. Student protests were the birthplace of groups such as the Maoist-Catholic Movimento Studentesco, whose members opposed an academic system that they considered capitalist oriented. Unemployed Italians grew restless and pressed for change, while workers, unhappy with factory conditions and low wages, turned to trade unions for help.

There were massive sudden strikes and riots in the fall of 1968, called the Hot Autumn. Some employers turned for help against the strikers to groups called *squadristi*, which were similar to the old Fascists.

At the same time, certain Socialist and Communist radicals began demanding violent change. Some of them, after being jailed on short-term sentences, formed connections with criminals. The result was the formation of a number of illegal political organizations.

The most ruthless and well-organized of these groups was the Red Brigades, founded in Milan by Renato Curcio in 1970. To fund their activities, the Red Brigades robbed banks and homes and kidnapped wealthy, influential people for ransom money. They terrorized the population by exploding firebombs in public places; they were known for mutilating their victims, and they committed their first known murder in 1974. But the Red Brigades were not alone in the field of terrorism. By 1976, 140 underground left-wing groups had been identified as responsible for more than 2,000 acts of terrorism each year.

In March 1978, the Red Brigades kidnapped Aldo Moro, a former prime minister and member of the Christian Democratic party who had antagonized them by being insufficiently radical. The Red Brigades demanded that certain terrorists be released from prison; when this demand was not met, they "tried" Moro, found him guilty, and executed him. Strong antiterrorist laws were put into effect shortly thereafter.

Since then, left-wing terrorist acts have been less of a problem, but right-wing terrorism has flared up now and then. In the 1990s, the Clean Hands campaign by the Italian judiciary resulted in a number of sensational charges against leading politicians. One former prime minister was convicted of corruption, and another was charged with conspiring with the Mafia in a murder. Overall, Italian politics continues to be turbulent.

Sculptor Marini, shown in his Milan studio with one of his works, is one of the 20th century's heirs to Italy's long tradition of artistic achievement.

6

Art and Culture

The Romans inherited highly sophisticated arts from ancient Greece and adapted them to their own purposes. The Romans were especially gifted as architects. They invented the brick-and-mortar method of building walls and perfected the use of the arch as a doorway, window, or roof support. They furnished private villas with central heating and indoor toilets. And their public structures—theaters, aqueducts, coliseums, circuses—were built so sturdily that many still stand today.

The Romans esteemed modesty and straightforward, honest artistic portrayals. Where a Greek portraitist would flatter his subject—perhaps giving a king flowing hair and a fierce expression—a Roman portraitist was likely not to improve on nature. The sitter's bald spot, bumpy nose, or jug ears were portrayed for all to see. Realistic portraits of the emperor, the representative of the Roman people, served as a kind of advertisement ensuring the government's modest, fair-minded policies.

Shortly after Constantine the Great declared Christianity the official religion in 313, a movement against the making of images arose. Iconoclasts, or "image breakers," tore down and destroyed sacred and royal portraits for fear that they might encourage pagan worship of idols. But one Christian leader, Justinian, the Byzantine emperor in Ravenna, realized that image making could be useful. Pictures could spread the gospel to pagans and illiterates. Ravenna's splendid mosaics (pictures made by piecing together tiny multicolored stones or tiles) of martyrs, the Virgin Mary, Christ, and Justinian and Empress Theodora must have moved even the firmest unbeliever.

Painting and Sculpture

The spreading Christian faith called for a new art that could communicate the essence of God and spirituality. Roman art, with its emphasis on man and nature, was inadequate. Gradually, the artists of the Middle Ages developed a highly stylized, abstract art that used symbolism more than images from the visible world. Italian artists, however, never entirely abandoned their classical past.

The sculpture of Nicola Pisano (who was active around 1258–78) for the Pisa baptistery (a building used for baptism) is striking for its similarities to Roman imperial art. Instead of being mystical and slender like the figures in most medieval art, Pisano's people are rounded, weighty, and clad in ample gowns. They stand in poses drawn from classical statuary and tomb sculpture.

Nicola's son, Giovanni, carried on the family tradition in Pisa, but the greatest artist of the period was a young shepherd named Giotto di Bondone. He lived around 1267–1337, and his work departed dramatically from medieval convention. Giotto introduced naturalism to painting by giving his figures recognizably human emotions. His scenes showing the lives of the Virgin Mary and Christ (painted in the Arena Chapel in Padua in 1305–6) and of St. Francis (in Assisi and Florence) convey a wide range of human feeling.

The painter Giotto lived and worked in Florence, where the Renaissance first flourished. Artists there were greatly influenced

A sculpture by Giovanni Pisano of Pisa shows the Three Kings (upper left) adoring the infant Jesus (upper right). Nearly all Italian art of the Middle Ages was Christian in subject.

by the scientific and mathematical theories being advanced at Italian universities and academies. By applying geometrical principles to painting they devised a pictorial perspective that made the flat surface of a painting look as if it had depth. They studied anatomy to learn how to draw accurate representations of human forms and movements. And in classical art and science they found a rich source of subject matter, technical knowledge, and style.

Most of the famous artists of the early Renaissance grew up or worked in Florence. These included the painter Masaccio, a revolutionary innovator with an austere, realistic style that influenced the rest of Renaissance painting; Fra Filippo Lippi, who made frescoes (wall paintings on wet plaster) and was known for his decorative style; Fra Angelico, who painted colorful, complex frescoes;

Domenico Ghirlandajo, another fresco painter who taught Michelangelo; Piero della Francesca, a mathematician whose precise works were organized according to geometric proportions; Paolo Uccello, who attempted a scientific approach to painting and was particularly concerned with the problems of perspective; and Sandro Botticelli, who was trained by Fra Filippo Lippi and who often painted classical subjects. There were also sculptors such as Lorenzo Ghiberti, creator of the Florentine Baptistery's bronze doors, his pupil Donato de Betto di Bardi Donatello, whose work is now regarded as the best embodiment of the Renaissance spirit, and the architect Alberti, who favored classical styles over contemporary religious symbolism.

The Bellini family, Jacopo and his sons Gentile and Giovanni, created the Renaissance style in Venice. But perhaps the greatest

A page from the notebooks of Leonardo da Vinci shows his design for chariots that would cut down enemies with whirling scythes.

painter in Venice during the Renaissance was Titian, who apprenticed with the Bellini sons. Titian developed an expressive style using many layers of brilliant colors; his reds were so notable that the word *titian* has come to mean a rich shade of auburn.

Michelangelo di Lodovico Buonarroti Simoni (1475–1564) was known for his *terribilita*—his awesome power. He viewed the human body as the perfect physical incarnation of god. At 25, he completed his first masterpiece, a marble sculpture called the Pietà that is now on display in St. Peter's Basilica in Rome. Pilgrims from all over the Christian world flocked to admire this gleaming figure of the Virgin Mary cradling the corpse of her crucified son. Michelangelo went on to create a statue of David in the Galleria dell'Accademia, Florence, and a marble statue of Moses in the Church of San Pietro in Vincoli, Rome. Although he did not consider himself a painter, the frescoes he painted in the Sistine Chapel are regarded as some of the world's masterworks.

The Milanese painter Michelangelo da Caravaggio (1573–1610) replaced Michelangelo's superhuman creations with peasantlike saints and biblical figures, shabbily clad men and women with reddened hands and feet. Caravaggio's brand of harsh realism met with strong objections from a public used to the heroic men and women of Michelangelo, and this artist's life was a sensational one, consisting of frequent drunken brawls, jail terms, and a murder he committed over a bet on a tennis match.

Renaissance painting gave way to a style called mannerism. The Mannerist painters experimented with elongated forms and harsh colors. One of the best representatives of mannerism was Jacopo Robusti Tintoretto, a Venetian known for his enormous, vivid canvases. Tintoretto worked so quickly that he was called Il Furioso, "the furious one."

The 17th century saw the emergence of the baroque style. It was highly theatrical, blending contrasting elements and making the viewer participate emotionally in the scene depicted. Many baro-

Amedeo Modigliani was one of Italy's best-known modern painters. His works are mostly portraits, characterized by elongated faces and simple, direct poses.

que artists were northern Italians, such as the architect Francesco Borromini, who was invited south to help refurbish Rome. But the artist who best represented the baroque style was the Naples-born Gian Lorenzo Bernini (1598–1680), a brilliant architect, painter, sculptor, and poet.

The next artistic style to take hold in Italy was neoclassicism, inspired by the excavation of ancient Roman towns such as Pompeii and Herculaneum and by the publication of prints of Roman antiquities. By this time, however, Italy had relinquished its role as leader in the arts. Neither during the Risorgimento nor in the years immediately after it did Italian artists gain international stature. Shortly before World War I, however, a new art movement called futurism sprang up. Its leaders—Umberto Boccioni, Carlo Carra, Giacomo Balla—painted bold, active compositions intended to glorify speed, industry, and machinery.

The metaphysical school of painters, including Giorgio De Chirico and Giorgio Morandi, responded to futurism with paintings characterized by haunting stillness. Amedeo Modigliani (1884–1920), who belonged to neither of these groups, drew inspiration from African tribal art. In more recent years the work of the sculptors Marino Marini (1901–66) and Giacomo Manzu (b. 1908) has been widely appreciated, but Italy is not presently regarded as being at the forefront of the fine arts.

Music

The language of music is Italian. The word *scale* comes from *scala*, meaning step, and the words used by composers to indicate directions to the conductor or the performers—such as *andante, allegro,* and *presto*—are Italian. The history of Western music began in the 6th century with Pope Gregory I, who developed the religious music known as the Gregorian chant. In the 11th century, Guido d'Arezzo standardized musical notation.

Music filled the medieval and Renaissance worlds. Monks chanted daily prayers and sang the Mass, with sections performed by soloists. Instrumentalists played for competitions and other entertainments. Composers such as Giovanni da Cascia and Jacopo da Bologna related stories of love and heroism in madrigals, a form developed further in the 14th century by Francesco Landini. Religious music reached new heights with Giovanni Pierluigi da Palestrina, born about 1525, who wrote church music for 42 years.

But Italy's unique contribution to music is opera. This form, in which a dramatic work is set to music, is thought to have originated in Florence in the 16th century. The modern opera was developed by Claudio Monteverdi (1567–1643), only three of whose operas survive in complete form. His *Orfeo* (1607) is considered to be the first important opera. Monteverdi was followed by the prolific Alessandro Gaspare Scarlatti, who wrote 115 operas. Scarlatti's contemporary, Antonio Lucio Vivaldi, also composed operas but is perhaps best known for his violin concertos called *Le quattro stagioni* (The Four Seasons).

Milan's magnificent opera house, La Scala, opened in 1778 with a work by Antonio Salieri, the jealous rival of Wolfgang Amadeus Mozart. When Italian society was not knocking down La Scala's doors to hear the newest opera, it was bound to be flocking to hear the newest sensation, violinist and composer Niccolò Paganini (1782–1840). Paganini's wild appearance and extraordinarily skillful playing gave rise to rumors that he was in league with the devil. He displayed his diabolical skills on exquisite violins made in the city of Cremona: the Stradivarius, Guamerius, and Amati. Surviving examples of the Cremona violin makers' craft are among the most highly prized musical instruments in the world.

The 19th-century operatic composers Gioacchino Antonio Rossini and Gaetano Donizetti were succeeded by the greatest of all, Giuseppe Fortunino Francesco Verdi (1813–1901). Verdi's contribution to opera was his mixing of the songlike aria with passages of conversational declamation. His first success was *Nabucco* followed by *Aïda*, *Rigoletto*, *La Traviata*, and *Otello*. They remain popular to this day. A generation later, Giacomo Puccini composed *La Bohème* and *Madama Butterfly* and eventually became as popular as Verdi. The foremost Italian musical figure of the 20th century was the conductor Arturo Toscanini. He enjoyed a long career and made appearances all over the world until his death in 1957.

Literature

The literature of ancient Rome was written in Latin, the basis of the Romance languages of Western Europe. The most important poet during Republican times was Gaius Valerius Catullus (ca. 84–ca. 54 B.C.), who explored the meaning and nature of love. Horace (65–8 B.C.), the son of a former slave, wrote satires and odes in praise of Roman life under Augustus; a century later Juvenal (A.D. 60–140) bemoaned the apathy of a luxury-ridden Rome. The other writers of this period include Virgil (70–19 B.C.), whose *Aeneid* describes the experiences of a Trojan warrior after the fall of Troy; Livy (50

B.C.–A.D. 17), whose *History* narrates the story of Rome from its founding; and Ovid (43 B.C.–A.D. 17), best known for the mythological tales contained in his *Metamorphoses*.

Some of these texts have been lost, but many were preserved in the Middle Ages by monastic scribes. The church Christianized classical texts. For instance, Virgil's *Eclogues*, describing the dawn of a peaceful age following a child's birth, was interpreted as a prophecy of the birth of Christ. The primary medieval text was, of course, the Bible, standardized and translated into Latin in its entirety by St. Jerome during the 4th century. The Algerian-born St. Augustine added to the church literature with a volume of *Confessions*, recounting his spiritual development.

As early as the 1st century B.C., spoken Latin began to branch off from written Latin; bits of this spoken version were preserved in

Although only three of his operas survive, the 17th-century composer Claudio Monteverdi is regarded as the father of the modern opera.

Dante Alighieri, shown here in a 16th-century portrait, lived and wrote in the 13th century. His masterwork was his epic poem of Christian allegory, The Divine Comedy.

graffiti at Pompeii. One of the first to compose in the spoken language was St. Francis, who sometimes preached in his native Umbrian dialect and whose *Cantico di Frate Sole* (Canticle of Brother Sun) is written in Italian.

Toward the end of the 13th century, a lyrical and original literature called the *dolce stil nuovo*, or "sweet new style," was introduced by Guido Cavalcanti. In the next century, Dante Alighieri, Francesco Petrarca (called Petrarch), and Giovanni Boccaccio, all writing in the new style, established Tuscan (the dialect of the region around Florence) as the language of Italian literature.

Dante Alighieri was born in Florence in 1265. An active member of the Guelfs, he felt that the pope should abandon worldly ambition and concern himself exclusively with the affairs of the church. When the Ghibellines came into power, Dante was exiled from Florence, to be burned at the stake if he returned. He spent most of his remaining days in Verona and Ravenna. His masterpiece, the *Commedia* (also known as *The Divine Comedy*) was written in the vernacular. Its three books, *Inferno*, *Purgatorio*, and *Paradiso*, describe the poet's imaginary voyages through hell, purgatory, and heaven.

Petrarch (1304–74) was a poet and scholar who, along with Boccaccio, is credited with creating the Italian Renaissance in literature. A concern with nature, psychological conflict, and classical themes characterized his work. He developed and perfected the sonnet form of poetry. Giovanni Boccaccio (1313–75) was a prolific poet, whose best-known work is a collection of stories and tales called the *Decameron* (which means "ten days"). It is about 10 young Florentine nobles who retreat from the city to escape the Black Death. To pass the time, they tell stories, folktales, fairy tales, and anecdotes.

The 15th century produced a new flock of Italian writers such as Baldassare Castiglione (1478–1529), author of *Book of the Courtier*, in which a party of noblemen and women describe the perfect, well-rounded man—the type of person that came to be called the Renaissance man. Ludovico Ariosto wrote an epic poem called *Orlando Furioso* that was popular all over Europe. Around 1515, Niccolò Machiavelli wrote *The Prince*, a look at how a ruler may wield power without regard for ethics.

From the 16th to the 18th century, a highly popular theatrical form called the *commedia dell'arte* flourished. Actors donned masks and improvised dialogue for stock characters: the lovers, the foolish rich man, the braggart, the joker. Carlo Goldoni (1707–93) transformed *commedia* into a sophisticated comic form in plays such as *The Mistress of the Inn* and *The Fan*.

The growth of nationalism in the 19th century was reflected in historical novels such as Alessandro Manzoni's *I promessi sposi* (The Betrothed). The realistic and practical writers of the 1870s, called the *veristi* ("the true ones"), were led by Gabriele D'Annunzio, an early supporter of fascism. These writers told regional stories and explored the psychology of their characters.

During the early 20th century three Italian writers, each working in a different genre, achieved particular renown. Italo Svevo wrote novels concerned with profound psychological self-examinations. Luigi Pirandello was a novelist and short story writer who was best known for his plays, including the classic *Six Characters in Search of an Author*. The poet Eugenio Montale helped define and establish modern Italian poetry.

At the same time, Filippo T. Marinetti founded the Futurist literary movement, but it became harder for writers to express themselves when Mussolini gained power. He considered litera-ture a frivolous pursuit unless it served the state. Authors who questioned his policies risked being silenced, exiled, jailed, and sometimes—as in the instance of the Communist leader Antonio Gramsci—killed. Mussolini revived a literary association called the Royal Academy of Italy to preserve "our national character accord-ing to the genius and the tradition of the race." Its members included D'Annunzio, Marinetti, and Pirandello.

Neorealism emerged as the most vital literary form during the period after World War II. The subject matter of neorealism is drawn from the rich diversity of modern Italy; characters are drawn from the peasants and workers of the lower classes. The Neorealists made the first new effort since Dante to give written expression to the Italian spoken language.

Italy has produced an impressive body of literature in the 20th century. Its leading writers have included Alberto Moravia, whose works depict decadence and moral emptiness among the middle classes; Primo Levi, a Jewish Holocaust survivor whose memoirs are noted for their compassion; and Italo Calvino, a writer fascinated with fantasy and puzzles. Journalist Oriana Fallaci made an international reputation by interviewing world leaders and publishing books such as *A Man* and *Interview with History*. The feminist writer Dacia Maraini alternates among essays, novels, plays, and poems but always concerns herself with women's lives in contemporary society.

Movies

In the late 1940s, Italy became an important force in the world of international films with movies characterized by neorealism, which reacted to the slick style of Hollywood movies with political points of view and harsh, grainy photography. Roberto Rossellini's *Rome,*

Film director Federico Fellini put Italy in the forefront of international cinema with such movies as La strada *and* La dolce vita.

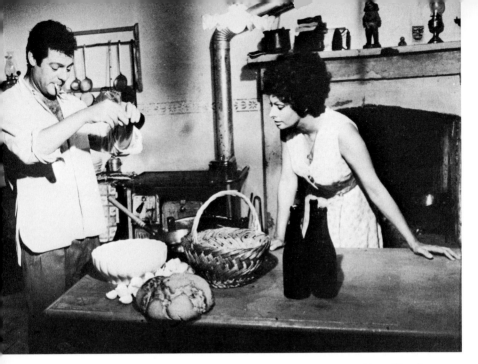

Marcello Mastroianni and Sophia Loren, shown here during the making of Sunflower *in 1969, are two Italian film stars who have achieved international renown.*

Open City (1945) featured nonactors alongside professionals and used documentary effects to portray working-class life under Nazi occupation. *The Bicycle Thief* (1947) by Vittorio De Sica is about a man whose livelihood is threatened when his only means of transportation, a bicycle, is stolen.

In the 1950s the *auteur* style of filmmaking emerged, in which the director develops a distinct, recognizable style, as in the case of Federico Fellini. Fellini's *La strada* (1954) and *Nights of Cabiria* (1956) established his style and earned him renown in international film circles. His three-hour *La dolce vita* (1959) ironically depicted "the sweet life" from the point of view of a playboy gossip columnist. Fellini's films became increasingly more symbolic and autobiographical with such explorations of the artistic temperament as *8 1/2* (1963) and *Juliet of the Spirits* (1965).

Michelangelo Antonioni, another filmmaker of the auteur school, began his career working with Roberto Rossellini and went on to develop his own style. His generally slow-moving films express the frustration and emptiness of modern culture, particularly with the rapid expansion of technology in postwar northern Italy. *L'avventura* (1959), *La notte* (1960), and *L'Eclisse* (1962) form a film trilogy concerned with this issue. Antonioni later made English-language films, including *Blow-Up* (1966) and *The Passenger* (1975).

Filmmaker Luchino Visconti began his career with *Ossessione* (1942), an early neorealist film. By the 1970s, however, he had turned to making opulent, beautiful movies such as *Death in Venice* (1971) and *Conversation Piece* (1975). Lina Wertmuller achieved great success in the 1970s with her politically charged sex farces *Love and Anarchy* (1972) and *The Seduction of Mimi* (1973). In 1975 she turned her attention to the horror of the Nazi occupation in World War II in *Seven Beauties*.

Italy has also produced internationally renowned film actors, such as Rudolph Valentino, one of the most popular silent-screen stars; Marcello Mastroianni, who appeared in many of Fellini's films; Sophia Loren, who had a popular career in Hollywood films in addition to appearing in some Italian movies; and Anna Magnani, who starred in *Rome, Open City* and later won awards for her work in the Hollywood version of Tennessee Williams's *Rose Tattoo*.

Pisa's Leaning Tower—a bell tower built in 1174—is one of many attractions that draw visitors to Italy from all over the world. Tourism is an important segment of the economy.

7

Economy and Society

Industrialization came late to Italy. During the second half of the 19th century, Italy was still an underdeveloped agricultural nation, resistant to change. Its few factories were mostly textile concerns. In the last decade of the century a number of new companies started up, among them the Olivetti typewriter company, the Fiat automobile company, and the Pirelli rubber company.

Mussolini set up Fascist agencies to represent employers and workers and to control the nation's industries. The Corporate State was dismantled at the end of World War II, but the new government retained Mussolini's notion of a government-directed economic plan. Italy came out of the war suffering from high inflation, a destabilized currency, and dire poverty in the south. In the next decades, Italy made leaps and bounds toward recovery, boosted by U.S. aid, the discovery of natural gas in the north and oil in Abruzzi and Sicily, and aggressive economic programs. By 1950, the lira had become one of Europe's most stable currencies.

In the mid-1950s, finance minister Ezio Vanoni embarked on a 10-year plan to reduce unemployment through government investment in troubled regions. Agencies were organized to attract business and industry to cities. The new programs also sponsored the drainage of marshlands for growing cotton and other crops. Between 1951 and 1978 a total of $27.7 billion was spent on the south in the form of loans and grants.

From the mid-1950s through the 1960s, unemployment dropped dramatically while wages rose. During the 1960s, Italy's gross domestic product (GDP)—the total value of goods and services produced—showed one of the highest growth rates in the world.

But in the 1980s and 1990s economic problems reappeared. In the mid-1990s, the unemployment rate hovered around 12 percent, too high for comfort. In addition, the growth rate of the GDP dipped to just 0.7 percent per year in 1996, one of the lowest rates among the industrialized economies. Italy's economy is hampered by the fact that most industrial raw materials have to be imported, as does the majority of the country's energy supply. Regional differences also persist. In spite of the government's investment in the south, that region remains largely undeveloped by industry, with poorer living conditions and more unemployment than in the north.

But Italy has many economic strengths to balance its difficulties. According to recent figures, Italy has become the fifth leading economic power in the world. It is a strong exporter, chiefly to its fellow members of the European Union but also to the United States and other nations. The country's leading exports include metals, textiles, clothing, industrial and agricultural machinery, vehicles and transportation equipment, and chemicals.

Agriculture

At one time the agricultural industry was Italy's primary employer. It now employs only 7 percent of the labor force. Crops are grown on 40 percent of the land. Most of this farmland is partitioned into small farms with an average area of 7.4 acres.

The main crop in the well-watered Po Plain is rice, while the south supports Mediterranean vegetation such as olive and citrus trees.

Spaghetti is made at a plant near Perugia; Italy produces and exports more than 420 varieties of pasta.

Wheat production has decreased and wheat imports have increased in recent years. Wheat-flour pasta is produced in more than 420 different shapes and sizes in factories in the provinces of Tuscany, Emilia, and Campania.

Many varieties of wine are produced in Italy. Among them are Chianti, Bardolino, Asti, Lambrusco, Marsala, Valpolicella, and the white wine of Orvieto. Italy is also among the leading producers of olive oil. Five different varieties are made, mostly in Umbria and Tuscany provinces. Other agricultural products include tobacco, corn (for animal feed), sugar beets, tomatoes, potatoes, and citrus fruits. Tobacco, food products, and beverages are also imported.

Half of the meat consumed in Italy is imported, although hogs and cattle are raised on the peninsula, sheep in Sardinia, and goats in the south. Italian cheeses made from cow's milk include Gorgonzola, mozzarella, provolone, and ricotta; sheep's milk is used to make pecorino cheese.

Energy and Mining

Italy is poor in mineral resources. Coal is excavated from Sardinian mines, and lignite (a form of coal) is mined in Tuscany, Basilicata, and Umbria. Modest amounts of sulfur are mined in Sicily. Italy also has deposits of iron ore, lead, pyrites, feldspar, and zinc concentrate.

By the mid-1990s, Italy ranked in the top ten in the world in primary energy consumption—that is, the use of petroleum, gas, and coal, as well as power from sources like nuclear and hydro-electric plants. Energy production, though, has not nearly kept pace with consumption. Imports, especially of petroleum, supply more than three-quarters of Italy's energy needs. Domestic petroleum production has grown sharply in recent decades but still falls far short of the quantity needed.

The Po Valley supplies most of Italy's natural gas, which is distributed through the country by means of 12,057 miles (19,400 kilometers) of pipeline. Thermal plants produce most of the nation's electrical power, though hydroelectric, nuclear, and geothermal sources are also important. In 1987, Italians voted to halt the building of nuclear power plants, and environmentalists hope to increase the country's use of renewable energy sources such as solar energy and hydroelectric power.

Manufacturing and Design

Milan and Turin are Italy's largest commercial centers, followed by Genoa, Naples, and Rome. Most Italian businesses are small; 60 percent of them have fewer than 100 employees. A few *condottieri*, or tycoons, dominate the commercial scene. Among these are the owners of the Ferruzi-Montedison chemical company and the Fiat car company.

Fiat, the country's leading automaker and a dominant force in the nation's industry, has weathered some setbacks in the past few decades. In the 1970s, for example, the company was targeted as an

The late Italian designer Gianni Versace gestures to the audience after unveiling his collection at a fashion show in Milan.

enemy by union militants who allied with left-wing extremists, and terrorists killed 4 executives and wounded 27 other company employees. Even so, Italy ranks among the world's ten largest producers of motor vehicles. By the mid-1990s, it was producing almost 1.5 million cars and 250,000 commercial vehicles per year.

Italy is also a world leader in textile manufacturing and fashion design. Designer names like Ferragamo, Gucci, Versace, Valentino, Fendi, Ferre, and Gigli have achieved international fame in recent decades. Georgio Armani is extremely popular in the United States, and outlets for the products of the Italian Benetton company are found worldwide.

Tourism

Italy's rich history and scenic beauty, together with the warm friendliness of its people, attract more tourists every year than any other European country. In 1996 more than 55 million foreigners visited Italy. Most came from Switzerland, Germany, France, Austria, the United States, the United Kingdom, Japan, and the Netherlands.

Culture and Society

Today's Italians have inherited from a mix of cultures that came about after the native populations were invaded by Germanic, Frankish, and Slavic tribes. The vast majority of them speak the Italian language and are descended from the same Italian stock, but centuries of disunification have created distinct regional cultures.

Much of Italian social life involves the extended family—grandparents, aunts and uncles, and cousins as well as parents and siblings—which often celebrates festive occasions with shared meals.

In general, the people of northern Italy consider themselves to be more cosmopolitan than their southern counterparts. Indeed, the north has more access to the rest of Europe, better industrialization, and better schools, whereas the south remains a relatively poor and agricultural region.

The people in some parts of the country speak their own dialects or languages closely related to Italian. For instance, Sardinian is spoken on the island of Sardinia; the Slavic community near the Slovenian border speaks Friulian; and the German-Italian community near the Austrian border speaks a dialect called Ladin. The German Italians are the largest minority group in Italy (making up about .25 percent of the population), followed by the Slavic Italians (.05 percent), the French Italians, and in the south some Greek, Albanian, and Serbo-Croatian communities.

The Italian social structure has always been based on the extended family. Children are most often raised in the company of parents and grandparents, cousins, and aunts and uncles. It is not unusual for Italians to live and work their whole life in the same region, close to the members of their extended families. Since World War II, however, this conventional way of life has changed somewhat. Educated, urban Italians have created an internationalist café society, which favors cosmopolitanism over provincialism—but most of the population still prefers traditional culture and custom.

One area that remains largely traditional is religion. Roman Catholicism was the state religion for many centuries, and Rome is the spiritual center of the worldwide Catholic church. Today, about 83 percent of Italians are Roman Catholic. Another 16 percent are either nonreligious or atheists, and the remaining 1 percent are mostly Protestants, Muslims, and Jews.

The international holiday May Day (also called Labor Day) is officially celebrated every year, and many of Italy's national holidays are observations of traditional Roman Catholic feast days: Epiphany, Easter Monday, Ascension of the Lord, Assumption of Mary, All Saints' Day, Immaculate Conception, Christmas, St. Joseph's Day, and St. Stephen's Day. Two others—Liberation Day on April 25 and the Festival of the Tricolor on May 12—are nationalistic in spirit. Italians also celebrate New Year's Day on January 1.

Italian women were given the right to vote in 1945. Today, more than one-third of Italian women are in the labor force, but they still do not earn what men earn. Since 1978, women in Italy have been able to obtain legal abortions. The Italian government pays 80 percent of a woman's maternity benefits and allows her to take as much as five months' paid maternity leave.

Government Services

Next to Germany, Italy has the most extensive and best-run transportation system in Europe, with 189,800 miles (305,388 kilometers) of heavily traveled roadways and 11,784 miles (18,961 kilometers) of railway tracks. Many other public services, however, are notoriously bad. Postal service can be extremely slow and has often been interrupted by wildcat strikes. Tax evasion and a lively black market deprive the government of much of the revenue needed to provide better services.

The Istituto Nazionale Previdenza Sociale provides unemployment, pension, and disability benefits. A national health plan instituted in 1980 guaranteed full medical coverage, with only minimal fees charged to the patient for some examination and treatment costs. In recent years, however, under pressure to meet European Union standards, the government has begun cutting back on social welfare programs, including pensions and health care benefits.

Although the country has more than 200,000 physicians, medical staff and facilities are poorly distributed outside the major cities. The biggest health problems are typical of industrialized nations:

Cancer and heart disease are the main causes of death. AIDS is a continuing problem. Italy has the third highest reported number of AIDS cases in Europe, after France and Spain; more than 38,000 cases had been reported by 1997.

School is compulsory for children between the ages of 6 and 14. Italy has the best teacher-pupil ratio (that is, the smallest classes) in Europe. The literacy rate is 98 percent; this means that nearly all adults can read and write. Colleges and universities are extremely overcrowded, in part because students tend to prolong their education because few job prospects await them upon graduation. The University of Bologna has about 60,000 students and the Univer-

In 1974, Italian women marched through the streets of Rome, demanding the passage of laws strengthening women's rights. As in many industrialized nations, women have entered the labor force in large numbers, but their pay is generally less than that of their male co-workers.

Schoolchildren in Rome attend a sports event. Children are required to attend school through age 14. More than 90 percent of Italian children go to public school.

sity of Rome has about 170,000, but the high cost of education limits university enrollment. Only about 4 percent of Italians have a degree from a university.

Media

The biggest national daily newspapers in Italy are the *Corriere della Sera*, which is published in Milan and Rome; *Il Giorno* in Milan; *La Repubblica* in Rome; and *La Stampa* in Turin. Because of the high cost of materials and labor, daily newspapers depend on contributions from corporations and big political parties in return for allowing these sources to control the papers' contents to some extent.

Until 1975 the government had monopolized public communications with an organization called Radiotelevisione Italiana (RAI). In 1975 the courts declared that all citizens had the right to free local information and created laws to ensure a measure of political independence in radio and television. Since then, 12 national and more than 450 local private television stations have begun broadcasting. A thousand local private radio stations have also commenced operations.

Italian Prime Minister Romano Prodi asks the Chamber of Deputies for a vote of confidence in his center-left coalition government in October, 1997.

8

Politics and the Law

Italy's form of government is parliamentary. The parliament has two chambers: the Senate, with 315 members, and the Chamber of Deputies, with 630. Male and female citizens age 18 and above are eligible to vote for deputies; those age 25 and above may vote for senators. Officials elected to the chambers serve five-year terms.

Under the 1948 constitution, the parliament elects a president who heads the republic for seven years. To win the election, the candidate must receive a two-thirds majority on one of the first three ballots or a plurality—more votes than any opponent—on a later ballot. The president nominates a prime minister who in turn appoints a council of ministers. They must be approved by both houses.

The Italian expression "the less government the better" reflects a long heritage of resistance to control by a centralized government. Federal policy frequently meets local opposition, and regional interests are commonly placed above national ones.

Several major parties dominate the political scene, and many of these are split further into smaller factions. Representatives from at least nine different parties usually hold seats in the Chamber of

Deputies. Thus, candidates seeking high office must form coalitions with several other parties in order to win an electoral majority. This confusion exasperates many Italians. Nonetheless, they are strongly committed to the democratic process, and most of them turn out at the polls on election day.

The country's first nationwide election, on April 18, 1948, gave the Christian Democrats—the Partito Democrazia Cristiana, or DC—an absolute majority. In the following decades, the DC continued to dominate, essentially monopolizing national power until the 1980s. The DC was closely linked to Roman Catholicism and was sometimes called the secular arm of the church. This association and the practice of rewarding supporters with government jobs helped the DC build a firm and lasting base of power.

Next to the DC, the Italian Communist party became the strongest party in Italy. It was, in fact, the largest and most powerful Communist party in the West. Its influence grew out of its role in the pro-Allied resistance during World War II, when Antonio Gramsci and Palmiro Togliatti were its leaders. In the 1970s the party gained new members by breaking with the Soviet Union and by softening its insistence on strict loyalty to a party line.

By the 1980s, however, other parties began to break through to national prominence. A Republican became prime minister in 1981, followed by a Socialist, Bettino Craxi, in 1983. Craxi served a longer term than any other prime minister had since the war. Craxi's government fell in 1987 after he antagonized the United States by releasing an Arab terrorist who had murdered an American aboard the cruise ship *Achille Lauro*.

In 1987, Christian Democrats regained the office of prime minister and remained in power until 1992. Soon after that, however, Italian political affairs entered an extraordinary time of turmoil. Severe scandals rocked the government, many of them the result of a "Clean Hands" initiative by Italian magistrates. Many political leaders were accused of corruption and of protecting the Mafia—or

Bettino Craxi, shown here addressing a meeting of the Italian Socialist party in 1978, served as Italy's first Socialist prime minister in the 1980s. In 1994 he was convicted of corruption.

even of active involvement in Mafia affairs. Craxi, convicted of corruption in 1994, fled the country. Giulio Andreotti, a seven-time DC prime minister, was tried for Mafia connections and for ordering the murder of a journalist. In the midst of the scandals, in 1993, voters approved a shift in the electoral system from proportional representation to largely majoritarian representation.

St. Peter's Square in Vatican City occasionally overflows with hundreds of thousands of people who come to hear papal pronouncements or to celebrate the Mass on holidays. Although no longer an official state religion, the Roman Catholic church remains influential in Italian life and politics.

Because of these developments, major changes took place in the national political landscape. The DC, so long the dominant party, effectively dissolved. The Communist party was succeeded by the Democratic Party of the Left. The Greens, a party dedicated to environmental issues, had entered the national scene in 1987, and now more new parties emerged.

By the mid-1990s, two large coalitions were developing: a center-right coalition called the Freedom Alliance, and a center-left group of parties known as the Olive Tree. Silvio Berlusconi of the Freedom Alliance served briefly as prime minister in the middle of the decade. Then, in 1996, Romano Prodi of the Olive Tree became prime minister. This political division into two major blocs represented a major change from the complicated multiparty politics of the previous decades.

Besides the political parties and coalitions, Italian political life is influenced by the three major trade union confederations and by associations of manufacturers, farmers, and merchants that act as pressure groups. The Roman Catholic church, though no longer as prominent in politics as it was during the long reign of the DC, remains an important influence. It is both an internationally recognized power and the country's one historically stable institution.

Church

Vatican City is the smallest city-state in the world, measuring only 108.7 acres. It has a population of about 730, including 70 Swiss guards. The Polish pope, John Paul II (born Karol Wojtyla), is the first non-Italian pope since the Dutch pope, Adrian VI, in 1522.

Although 83 percent of Italians call themselves Catholics, only 30 percent attend mass regularly. The status of the church has changed in recent years. A 1984 agreement between the Vatican and the Italian government revised Mussolini's Lateran Treaty and ruled that Roman Catholicism was no longer the state religion. Beginning in 1990, Italian taxpayers had the right to decide what portion, if any, of their tax money would help pay clerical salaries.

Judicial System

Italy's judicial system originated in the 13th century with cases of religious heresy brought before the Inquisition. It was standardized by Napoleon in the 19th century and codified by Mussolini in the 20th. Like every other European nation except the United Kingdom and Ireland, Italy uses the inquisitorial court procedure.

Under this system, most crimes are investigated by a judge, called the *pretore*, or praetor. Although serious cases are tried by six-member juries in consultation with two judges, most cases are heard by a judge alone. The judge questions witnesses, determines the defendant's guilt or innocence, and imposes the sentence.

In past years, the court system has proved vulnerable to abuses, especially to corruption through bribery. The system has also been slow, plagued by a severe shortage of staff and equipment. At times, it has let innocent people languish in jail for months. To help resolve such problems, the parliament instituted several important reforms in 1989. Some court procedures were modified to resemble those in the United States. Lawyers were allowed to question witnesses without a mediating judge, and defendants were allowed to submit a type of plea bargain. As a result, settlements between defendant and accuser could be reached before they set foot in court.

Despite its many critics, the Italian judicial system displayed its strengths in the 1990s with the so-called Clean Hands campaign. One magistrate, Antonio Di Pietro, had been compiling a computerized record of bribes and kickbacks. When an official in Milan was brought before Di Pietro in 1992 on charges of corruption, the judge soon realized the case had wide implications. Rapidly the investigation spread to other cities and to the national government, resulting in a nationwide scandal that implicated thousands of politicians, administrators, and business people. The prime minister,

Bettino Craxi, was charged and convicted, and in the 1994 elections the voters replaced 70 percent of the members of parliament.

By flexing its muscles in the Clean Hands campaign, the Italian judiciary demonstrated its potential for vigor and independence. Many observers hoped it would continue to develop these qualities in the 21st century.

A former appeals court judge, Corrado Carnevale, makes his way through reporters as he prepares to testify at the trial of former Prime Minister Giulio Andreotti, accused of Mafia-related corruption.

Organized Crime

The Mafia originated in Sicily after 1812. At that time, absentee landlords and barons of the huge estates in western Sicily hired intermediaries and private militias to terrorize farm laborers and maintain absolute control over them. These thugs organized themselves into bandit gangs and controlled all affairs within an area of Sicily that they carved out for themselves.

Mafia is now a general term describing organized crime groups operating in southern Italy, the leaders of which exercise control over others through intimidation and fear. This leader, called a *capo*, must be willing to kill all those who get in his way, whether they are trade unionists, law enforcement officials, judges, politicians, or rival *mafiosi*. A low-ranking member who oversteps his bounds is said to have taken "a step longer than his leg" and is swiftly punished.

Members avoid the word *Mafia*, preferring instead to call their organization the Cosa Nostra. The various groups have built networks with other organized crime families, government officials, and the police. Their wealth comes from kickbacks for contracts on public works, from protection money forked over by local merchants, from underhanded real estate deals, and, more recently, from the highly lucrative drug trade. There are three major organized crime groups in Italy—the Mafia in western Sicily, the Camorra in the western coastal province of Campania, and the 'Ndrangheta in the southern province of Calabria.

Recent years have brought serious attempts at a crackdown on the Mafia and related groups. A major trial in December 1987 sent 338 of the 452 defendants to prison. A later trial in Turin resulted in the conviction of another 130 members of organized crime families. By the beginning of the 1990s, a "high commissioner against the Mafia" had been appointed by the parliament, and in 1993 the Mafia's "boss of bosses," Totò Riina, was captured along with his chief lieutenant.

The Clean Hands campaign succeeded in indicting quite a few additional Mafia bosses, and the revealed connections between the Mafia and the political elite—long suspected but never before proved—appalled many Italian citizens. Even in Sicily, voters rejected some local candidates who were believed to have Mafia ties. Still, the Mafia has deep roots, and few observers think it will wither away.

Confronted by the economic, political, and environmental challenges of the contemporary world, many Italians draw strength from traditional sources: church, regional identity, and especially family life.

9

Italy's Future

For a long time Italy lagged behind other Western nations. When other countries were assuming their present-day political and geographical identities, Italy remained fragmented. Powers such as France and England were vying for global political dominance while Italy was still taking its first steps toward unification. Even after Garibaldi's triumphant march, foreigners continued to control Italy's destiny. And in the dawn of the 20th century, Italians clung stubbornly to an economy based on agriculture and moved only reluctantly and late toward industrialization.

Italy's hold on the past may have hindered the country's growth at times, but it also benefits the Italians in many ways. Italian families are far more stable than families in other countries. Parents continue to live in their ancestral towns and bring up their children in the old traditions. The family remains a constant and reliable source of shelter and strength in the face of difficulties presented by the modern world. Members of the younger generation, many of them little inclined toward the restlessness felt by young people in other Western cultures, still look to their elders for advice and hope to learn from their experience.

Once it was the Etruscan town of Faesulae, then the Roman city of Florentia, and later still a Renaissance city-state. The Florence of today, like all of Italy, refuses to be simply a shrine to the past and is seeking to shape an international economic and political role for the future.

Italians maintain age-old customs. For example, in every community, from small towns like Assisi to cities as large as Rome, young people and old partake in the daily *passeggiata*—the customary evening stroll. As they walk along the main street connecting the city's *piazzas* (squares), neighbors greet one another, discuss the day's events, argue politics, trade sports stories, and maybe stop in a café to share a glass of wine. The outdoor marketplaces are alive with people haggling over prices and joking with one another. The dinner table also provides a gathering place for animated conversation, and great care is devoted to the selection and preparation of good food and drink.

The work of skilled artisans commands enormous respect. Italians honor the effort that goes into making an exquisite lace tablecloth, crafting a fine pair of shoes, or building a sturdy cabinet. Furnishings, even in the humblest of homes, are likely to be meticulously made and passed down through the generations. The Italian flair for style and grace carries over into the contemporary design of such products as automobiles and clothing.

For all the charm of Italy's traditions and the admirable efforts of its people to preserve their noble past, Italians endanger themselves by failing to keep up with other modernizing nations. Journalist Luigi Barzini hints at this when he urges that Italy should change its image "from that of a picturesque, possibly lovable . . . country, a tourists' paradise, to one that should be taken seriously, treated with respect, and possibly even feared."

Though weary of the constant turmoil within their borders, Italians have adjusted to the volatile nature of their political and social institutions. The country manages to function, even to prosper, in spite of the often sorry state of public services, an overburdened legal system, and ongoing political scandal.

The increasing unification of Europe through the European Union (EU) may draw Italy into a larger role in international affairs. Compared to other Europeans, Italians have shown little resistance to the EU's growing impact. One reason is that the Italians, a people late to unify as a single nation, have never had as fierce an allegiance to their own government as the French, say, or the British. But will Italian politics prove any less turbulent when some of the policies originate at the EU headquarters in Brussels?

Over the centuries, Italy has shown itself able to adapt to an enormous variety of changing circumstances. Its future holds great challenges and enormous promise.

GLOSSARY

Carbonari   Members of a secret society founded in the 19th century. It supported Italy's independence and the formation of a constitutional government.

commedia dell'arte   Improvisational comic theater by masked actors playing traditional roles. This form of theater flourished from the 16th through the 18th centuries.

condottieri   Business tycoons who run the large Italian industries.

dolce stil nuovo   The "sweet new style" in literature introduced by Guido Cavalcanti toward the end of the 13th century. Writers in the style, such as Dante, Petrarch, and Boccaccio, replaced literary Latin with the vernacular, or spoken Italian.

Guelfs and Ghibellines   Two rival factions active in Italy from the 13th to the 15th century. The Guelfs supported the papacy and the Ghibellines favored the empire.

Mafia   A general name for organized crime groups in Italy including the Mafia of Sicily, the Camorra of Campania, and the 'Ndrangheta of Calabria.

passeggiata   The evening stroll taken by Italians in every town.

piazza   A city square; the center of a small town.

pretore   A judge.

Risorgimento   The "resurgence," a period of nationalism in the mid- and late 19th century that led to the formation of a united, independent Italy.

risotto   Rice, grown in the fertile Po Valley of northern Italy and a basic ingredient of northern Italian dishes.

INDEX

128